'Jeffreys has achieved what no other writer on English wine has managed to do. He's written a book that is impossible to put down. It's as much fun as a novel. [...] It's a brilliant read, written with immaculate style and poise, it's funny, witty, spiky, irreverent, sometimes even a little shocking. [...] English wine will never taste the same again.' Tamlyn Currin, www.jancisrobinson.com

'Jeffreys' entertaining, accessible and skilfully paced book helps us relish the English countryside's delicious new calling.' Andrew Jefford, wine writer and author of *Drinking with the Valkyries*

'A fine history. [...] Witty and erudite.' Adam Lechmere, *Club Oenologique*

'A fascinating and superbly told adventure.' John Clarke

'It's a good read [...] in a warts-and-all style.' Stephen Skelton, X

'Rather like a novel you can't put down.' David Crossley, Wide World of Wine

'[Jeffreys has] done a great job of highlighting the peculiar Britishness of the whole endeavour, and it's shot through with the wry humour that makes his writing so enjoyable.' Matt Walls, contributing editor *Decanter* magazine

'Jeffreys has written quite a fine book on a subject that's already been covered a great deal to date. He makes it fresh and vibrant through firsthand research and interviews as well as his many years as a drinks writer.' Miquel Hudin

'Jeffreys writes with gentle wit and an informal style that neither overplays the bonhomie nor condescends [...] well-paced, engaging and packed full of interest. It's also incredibly well researched and balanced. [...] For anyone – Brit or otherwise – who still needs convincing, *Vines in a Cold Climate* comes highly recommended.' Simon Woolf, *The Morning Claret*

VINES
IN A COLD
CLIMATE

The People Behind the English Wine Revolution

HENRY JEFFREYS

ALLEN&UNWIN

First published in hardback in Great Britain in 2023 by Allen & Unwin, an imprint of Atlantic Books Ltd.

This paperback edition first published in Great Britain in 2024 by Allen & Unwin.

Map illustration by Jeff Edwards

10 9 8 7 6 5 4 3 2 1

A CIP catalogue record for this book is available from the British Library.

Paperback ISBN: 978 1 83895 667 7
E-book ISBN: 978 1 83895 666 0

Printed and bound by CPI (UK) Ltd, Croydon CR0 4YY

Allen & Unwin
An imprint of Atlantic Books Ltd
Ormond House
26–27 Boswell Street
London
WC1N 3JZ

www.atlantic-books.co.uk

MIX
Paper | Supporting
responsible forestry
FSC® C171272

For Misti

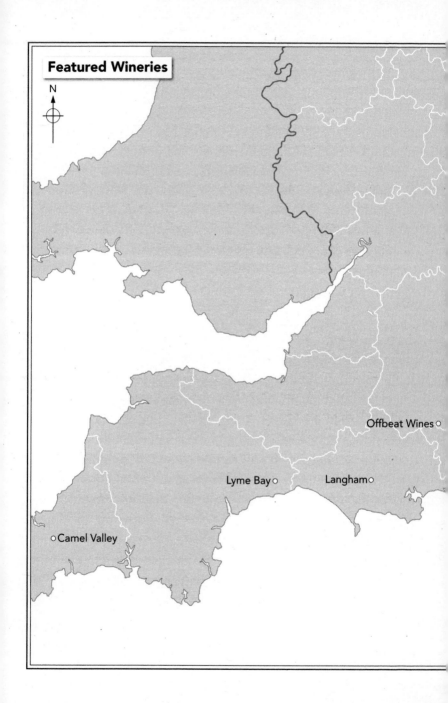

Featured Wineries

N

Offbeat Wines○

Lyme Bay○   Langham○

○Camel Valley

Flint○

○Gutter and Stars

Danbury Ridge
○○ New Hall

Hundred Hills
○
○
Harrow and Hope
○
Blackbook

Domaine Evremond
○
Westwell○   ○Simpsons
○Balfour

○Hattingley Valley
Gusbourne○
○Davenport   Chapel Down
Nyetimber   Ridgeview   Tillingham
○   ○   ○
Hambledon○   Plumpton College○
○
Breaky Bottom ○Rathfinny

| 0 | | 25 miles |
| 0 | | 50 kms |

CONTENTS

'The luscious clusters of the vine
Upon my mouth do crush their wine'

Andrew Marvell

INTRODUCTION

On a blustery, unseasonably cold May day in 2017, the cream of Britain's drinks press descended on a field just outside Faversham in Kent for a milestone event in the history of English wine. Taittinger was planting vines in southern England – and we had been invited to take part.

The week before, late spring frosts had damaged vines across the country. Some growers had lost 80 per cent of their crop. Combine that with all the uncertainty about the previous year's referendum result, in which Britain had voted by a narrow margin to leave the European Union, and you might say that Taittinger's timing could have been better.

The French company had bought the land in 2015, after years of rumours that Champagne houses were looking to make wine in southern England. It was followed, in 2017, by Pommery, which would become the first Champagne house to actually launch an English wine, made in conjunction with Hattingley Valley in Hampshire. Both were following in the footsteps of a lone winemaker from Champagne, Didier Pierson, who had beaten all the big boys to it when he planted vines in Hampshire in 2005 and began making sparkling wines under the Meonhill label (since bought by Hambledon).

To make high-quality sparkling wine by the *méthode champenoise*, you need grapes with high acidity. They need to be ripe, but not too ripe. With the climate in Champagne getting warmer, southern England is arguably the next best place on earth to grow suitable grapes. It even has chalky soil identical to that of Champagne.

On the day of our visit, we piled out of the buses from Ashford station at a nondescript, muddy field in what felt like the middle of nowhere. We had been warned to dress casually and to be 'prepared for the unpredictable British weather. The event is taking place in a field and we have very limited cover'. Many urban types had not heeded the advice, wearing smart shoes and even heels.

Shivering outside, we sipped tea to warm us up and then strode out somewhat gingerly into the field for the planting of the vines. The rain was horizontal, like you get on Scottish islands. Patrick McGrath a Master of Wine from Hatch Mansfield, Taittinger's UK distributor and partner in the venture, stood on a box and tried to make himself heard above the wind. Then it was the turn of Pierre-Emmanuel Taittinger from the family that owns the Champagne house, dressed up as an English gent in that charming way certain Frenchmen in the wine trade still do.

He insisted that bonds between Britain and France, and more specifically Champagne, would endure despite Brexit. Taittinger's Kent venture is named Domaine Evremond after Charles de Saint-Évremond, a French aristocratic exile in the court of Charles II who introduced the wines of Champagne to

England, where they were served at parties – some say orgies – attended by the king. For this service, Charles made Évremond governor of Duck Island in St James's Park, which came with a £300-a-year salary.

As the rain got heavier, the PR team cut the speeches short. We were handed ceremonial trowels, given vine cuttings and shown where to plant them. I sometimes wonder how mine is doing, hopefully thriving somewhere in the damp Kent soil.

Did I mention that it was really cold? Our job done, we hurried into the marquee. We couldn't taste wines from Domaine Evremond – they won't be released until 2024 at the earliest – so in a clever bit of publicity, Taittinger had invited other Kentish producers to show off their sparkling wines. There were wines from Chapel Down, England's largest producer, and Gusbourne, one of the country's most prestigious, as well as newer names like Squerryes (rhymes with cherries) alongside veterans like Biddenden. The quality was high, with none of the searing acidity that has sometimes characterised English wines in the past. Perhaps aware of the comparison, or as a bit of flattery, Taittinger did not offer its standard label, but instead brought out dozens of bottles of its £150 top-of-the-range Comtes de Champagne.

With everyone thoroughly refreshed, it was Taittinger's turn to speak again. He clearly, gloriously, had had no media training and rambled charmingly on subjects ranging from his recollection of English women encountered in his mis-spent youth – there's more than a touch of Évremond about Pierre-Emmanuel – to how Kent held some advantages over Champagne, not least the lack of unexploded World War One munitions in the vineyards.

3

With the rain, the sweaty marquee, the increasingly drunk guests and the risqué speech, the event had more in common with a British country wedding than an event put on by a French wine company. Taittinger had somehow contrived to make the most English day out possible for the launch of Domaine Evremond. And yet underneath all the fun, there was clearly a deadly commercial intent. Taittinger was investing millions in this.

Finally, Taittinger ended his speech with the thought that perhaps Domaine Evremond would one day attract tourists from France to England. 'There is this beautiful unexplored island off the coast of France,' he said. French people coming to taste English wines – now wouldn't that be something?

* * *

The money, the world-class wines, the slick PR – it's all a far cry from my first experience of English wine. That was at a wedding in the early 2000s at a country house in Suffolk that made its own wine. I can still remember the peculiar taste: initially quite sweet, then chalky, followed by masses and masses of acidity. This wasn't a German mouth-watering acidity like you get in a Mosel riesling. No, this was acidity so hard it reminded me of the stone floors of my boarding school. And there was no fruit at all. It didn't taste like any wine I'd encountered before.

After that, I had English wines occasionally but even the best had a similar lack of fruit and a hardness to them. The problem with England as a place to grow vines is not just that it's cold, but that it's also grey and wet. There's often not enough sunshine to ripen grapes properly and the damp makes them prey to rot. Except in exceptional years, it takes a lot of care to ripen classic French or German grape varieties in this climate.

So English growers planted varieties like müller-thurgau, designed to ripen reliably early in cooler climates. But the growers were often ill-prepared amateurs. Vines were planted in places prone to poor sunlight, bad drainage, frost traps and generally at the mercy of the elements. Winemakers over-compensated for lack of ripeness by adding sweetness in the form of sugar or unfermented grape juice, thereby yielding pale imitations of humdrum German wines. And who needs England's answer to Blue Nun?

English wines, with a few exceptions, were novelties, sold to holidaymakers in southern England. Or so I thought. In fact, there was a quiet revolution going on in the English countryside. It had been noticed by some pioneers that southern England's marginal climate was ideal for making sparkling wines and boasted chalky soil just like that in Champagne. An American couple with no winemaking experience, Stuart and Sandy Moss, planted the classic Champagne grapes – pinot noir, pinot meunier and chardonnay – at Nyetimber in West Sussex. 1992, their first vintage, won a gold medal at the International Wine and Spirits Competition.

I tasted my first English sparkling wine in the late noughties. It was made by Ridgeview, one of the country's largest producers, in East Sussex, and I liked it. Though the acidity was a little racy, it was clearly a very well-made wine. But it was another wine, also made by Ridgeview, a few years later that really changed my perceptions of English wine. It came from a tiny plot which has now been pulled up near Reading called Theale vineyard owned by the Laithwaite family, the people behind Britain's largest mail-order wine merchant. It was appley and rich, and much more delicious than any Champagne at the same price.

Others clearly thought so too. It became quite the thing to conduct blind tastings of the best sparkling wines, and the English ones often came out on top. The first was the so-called 'Judgement of Parsons Green' organised by wine consultant and Master of Wine Stephen Skelton in 2011 with a Ridgeview Grosvenor Blanc de Blancs 2007 emerging triumphant. It was followed by various other 'Judgements' culminating in 2016, when, at a competition in Paris judged by actual Frenchmen, a wine from Nyetimber beat France's finest. It was like Agincourt all over again. The newspapers had a field day.

But it wasn't just in the news pages that English sparkling wines were proving popular. They were also winning over drinkers. Bars like London's St Pancras Hotel began offering English wine as the house fizz, instead of non-vintage Champagne. Vineyard planting increased from around 1000 hectares in 2005 to nearly 4000 in 2018. Seventy per cent of this was used to make sparkling wine but the still wines were coming on rapidly too. Bacchus, one of those Germanic crosses with its crisp nettle, grapefruit and elderflower flavours, was touted as England's answer to sauvignon blanc. By 2010, wines like Chapel Down Flint Dry were ubiquitous in British supermarkets and people were opening them without that knowing wink that said 'It's English, can you believe it?'

Unlike the sparklers, however, it took me much longer to come round to the joys of England's still wines. In 2011, I started a wine column for *The Lady* magazine. The pay was terrible and erratic, but I stuck it out because it amused me to introduce myself at parties as the wine critic for *The Lady*; people would then treat me as if I had stepped out from the pages of P. G. Wodehouse. The editor, Rachel Johnson – sister of Boris – asked me to write something on English wines. Somewhat reluctantly, I agreed.

At the time there was a shop in London's Borough Market called the Wine Pantry that sold only English wines. My wife and I went down one day and met with Julia Stafford, an English wine evangelist who ran it. We tried bottle after bottle and I found several whites that really impressed me. Some were made from the usual crosses designed for cool climates, like bacchus and huxelrebe, but there were also French varieties like pinot blanc and pinot gris, and some of them were pretty good. My mind was beginning to open.

I started coming across landmark English still wines: an exceptional 2013 chardonnay from Gusbourne; a pinot noir from Bolney, the first English red where I actually wanted to finish the bottle; another chardonnay, this time from Kit's Coty, Chapel Down's single-vineyard premium range. And it wasn't just the French varieties – an ortega, a German grape, grown by Biddenden in Kent, tasted wonderfully distinctively English. Then there were organic blends from Davenport in Sussex.

With crippling frosts at the start of the growing season, 2017 was a terrible vintage for many growers but the following year was a corker, with some vineyards able to get their red grapes riper than ever before. Gusbourne in Kent made its best-ever pinot noir in 2018, one that amazed me with its ripeness and perfume. Here was a red that wasn't just good for England, it would have been notable if it came from France or Germany. Now there's even riper pinot being made by a vineyard in Essex called Danbury Ridge. Its 2020 vintage had a brightness of fruit that was positively Californian.

So how did we get, in 30 years, from boarding-school acidity to making pinot noir that gives Burgundy a run for its money? It's a fascinating story of triumph and disaster, full of larger-than-life characters with big ambitions. Most of the story takes place since the 1990s, but the roots of English winemaking stretch back to medieval times and perhaps even further to the Roman occupation.

This book is not a guide to English wine – your favourite producer may not even be mentioned. Very quickly in my research, I realised that there was too much going on to include everyone. Instead I've picked a small group of people and producers who are emblematic of the rapid changes within the industry. I have concentrated on southern England, as something like 90 per cent of the wine made in Britain today comes from grapes grown in this part of the country (though there are at least two excellent producers in Wales). There's a bibliography at the back for those who want to explore further.

My aim is to show how English wine went from a joke to world class in 30 years. There's no doubt that the changing climate has played a huge part in this story. Global warming has so far been good for English wine, though it hasn't all been positive. Warm winters followed by cold springs bring the risk of frosts like those which wreaked such havoc in 2017. If warming continues at the same rate, southern England could become too hot to make its new-found signature wines.

But this isn't a book about the weather, either. It's about cooperation and conflict, inspiration and perspiration, hope and doubt. Most of all, it's about a few determined, some might say bloody-minded, people. From the City types with nothing but a dream and a spare few million to a single mother working a few acres in Kent, it's the story of a handful of men and women who ignored the doubters like me and decided that not only could you make drinkable wine in England, you could make something truly world class. And all from our cold, damp climate.

CHAPTER 1

False starts

'The whole company said they never drank better foreign wine in their lives'

Samuel Pepys

Searching for evidence of historical winegrowing in England can take you to some funny places . . . like California. There's a wine made by Richard Grant, a Napa Valley producer, called Wrotham Clone Reserve Pinot Noir. It's named after an obscure version of the pinot noir grape that comes not from France, but from a village in Kent between Sevenoaks and Maidstone called Wrotham (pronounced 'root 'um') where it was found growing against a wall. All very mysterious.

The vine was discovered by an English wine pioneer called Edward Hyams, who was born in Stamford Hill in London in 1910. Photos show a man who in later life sported an enormous moustache like Georges Clemenceau. But don't let his Edwardian appearance fool you – there was something of the Tom Good from *The Good Life* about Hyams. A historian, sci-fi novelist, journalist, ecologist, horticulturist and pacifist, he was a founder member of the organic farming body the Soil Association. After the Second World War, in which, despite his pacifism, he served in the RAF and Royal Navy, he moved with his wife to Molash in Kent to pursue his interest in horticulture, particularly vine growing. The idea was to be fully self-sustainable. In his memoir *From the Wasteland*, published in 1950, he admitted that part of the reason for his interest in viticulture was so that they would be able to drink a litre of wine a day between them. Which sounds like as good a reason as any.

This was at a time when there had not been any commercial vineyards in the British Isles for 35 years. Not only did Hyams grow and attempt to make wine, with varying degrees of success, but he was a proselytizer for English wine, writing books and newspaper articles on the subject as well as being a regular voice on the BBC. He was always on the hunt for evidence of historic viticulture, and one day in the 1950s he struck gold when he discovered a strange grape variety growing wild in a churchyard in Wrotham.

That Wrotham vine is thought to have been around 200 years old when it was discovered. With its white-dappled leaves, the vine resembled pinot meunier (the name 'meunier' means 'miller' because the vine leaves were said to look like they had been dusted with flour).[1] How it got to Kent is a bit of a mystery. One theory is that it's the same variety as a vine identified by Sir Joseph Banks at Tortworth in Gloucestershire,

---

1   Though later DNA testing would show that Wrotham pinot is an early-ripening clone of pinot noir

known as Miller's Burgundy (named after a Mr Miller rather than because of its dusty leaves). Banks, who lived from 1743 to 1820, was a botanist and a member of the Royal Society. In a picture by Joshua Reynolds that sits in the National Portrait Gallery, he looks a dashing sort of fellow, every inch the dishy, romantic scientist of the popular imagination. Sadly in his later years he cut rather less of a dash, as he suffered so badly from gout that he struggled to walk and had to be wheeled around.

Two centuries later and with his flair for publicity, Hyams surmised that pinot noir was brought to Britain by the Romans, a claim that has since been repeated in much literature on the subject. Most books on English wine start with the Romans planting vines in England. It's a tantalising link, especially as one of the most popular grapes in England is called bacchus, after the Roman god of wine and merriment. Sadly, bacchus as a variety actually dates back just to 1933, comes from Germany and was only planted in England in the seventies.

Tacitus wrote of Britain that 'the sky is obscured by constant rain and cold, but it never gets bitterly cold.' Though he never actually visited, the Roman historian's assessment sounds pretty accurate. According to analysis from tree rings, it was likely that the climate in Roman Britain was about 1°C warmer than it is today. Hence it would have been perfectly possible for the Romans to grow grapes here – though if they did, no conclusive evidence survives. We know that plenty of wine was drunk in Roman Britain, as broken amphorae with wine residues attest, but it was most likely imported. Just as it is now, Britain was plugged into a sophisticated trade network where wine from warmer climates could be brought to these shores more easily than growing grapes on a damp, dark island. This would be a perennial problem for England's winemakers: why struggle to make what you can more easily import?

The balmy climate of Roman Britain didn't last. Around 400 AD it became colder and wetter. There's a theory that this cooler weather hastened the decline of the Western Roman Empire as northern tribes moved in to escape the cold. The climate change certainly ended any viticulture that was going on in Britain. By the 10th century, however, the climate had begun to warm up again. This was the start of the medieval warm period that would last until around 1300. The Domesday Book lists vineyards all over the south of England but particularly in East Anglia and the South West. Some of these were long-term enterprises, as Hugh Barty-King writes in *A Tradition of English Wine*: 'Many vineyards which had featured in the Domesday Survey were also being worked into the fourteenth century.' Oddly, vines were thin on the ground in the heartland of modern English wine, Kent and Sussex, perhaps because the areas were still heavily forested at the time.

According to Hyams,[2] England's vineyards were not of in-significant size, nor were they just the preserve of monasteries. The 12th century chronicler William of Malmesbury wrote of his home county of Wiltshire that: 'The vines are thicker, the grapes more plentiful and their flavour more delightful than in any other part of England. Those who drink this wine do not have to contort their lips because of the sharp and unpleasant taste, indeed it is little inferior to French wine in sweetness.' There's a detail of a grape harvest in a carving in Gloucester Cathedral, and there is evidence for grape growing in place names containing 'win', 'wyn', 'vyn', 'vin', 'vine' or 'vyne'.

---

2 *Dionysus* by Edward Hyams

Indeed, the city of Winchester in Hampshire may be named after the vine. The poet Robert of Gloucester wrote how, 'London is known for its shipping. Winchester for its wine.'

In 1154, King Henry II married Eleanor of Aquitaine. The couple did have a vineyard in England, at Windsor, not far from where the Laithwaite family has now planted grapes in Windsor Great Park, but the influx of wine from Bordeaux and the surrounding area which was now under English control was a blow to the home-grown product. Hyams wrote that, 'The infant English industry was overlaid at birth by its immensely vigorous Gallic mother.'[3] Now there's an image. Even after England lost control of Aquitaine following the end of the Hundred Years' War at the battle of Castillon in 1453, French imports continued. It wasn't just coming from France either – wines from southern Spain and the Canary Islands, Madeira, Portugal, Italy and Cyprus were common too.

In his diary, Samuel Pepys boasts of the variety of wines in his cellar in 17th century London: 'At this time I have two tierces [a small cask] of claret [red Bordeaux] – two quarter-casks of canary [wine from the Canary Islands, probably not dissimilar to sherry], and a smaller of sack [sherry] – a vessel of tent [red wine from Spain], another of Malaga, and another of white wine, all in my wine-cellar together – which I believe none of my friends of my name now alive ever had of his own at one time.' There were no English wines in his cellar but in his diary he did mention visiting Hatfield House, north of London, where Lord Salisbury had a vineyard. He also visited a vineyard

---

3   From *Dionysus*. And what a vigorous mother she was. During the 14th century, it's estimated that something like the equivalent of 50 million bottles of wine would be shipped from Bordeaux to England every year at a time when the population was around 5 million

belonging to a Colonel Blunt near Blackheath, though sadly he didn't comment on its merit. But another diarist, John Evelyn, tried Blunt's wine, and pronounced it 'good for little'.

Like Pepys, Evelyn was a noted drinks enthusiast, a member of the Royal Society and a cider maker. Evelyn also wrote a book called *Pomona*, aimed at landowners, which argued that rather than make or import wine, the English should drink high-class cider instead. He wrote: 'Our design is relieving the want of wine, by a succedaneum [substitution] of Cider.' It wasn't for another three years, in 1667, that Pepys first recorded his thoughts on English wine, when he wrote about one made by Admiral Sir William Batten from grapes grown in his garden at Walthamstow, then a village outside London. Pepys wrote of trying 'a bottle or two of his own [Batten's] last year's wine, growing at Walthamstow; then the whole company said they never drank better foreign wine in their lives'.

Despite the inclement weather of this period (1500–1700, better known today as the Little Ice Age), English viticulture was gathering pace. In 1666, John Rose published a book championing native wines called *The English Vineyard Vindicated*. He was Charles II's personal gardener; there's an amusing-looking picture in the National Trust collection of Rose on his knees presenting a pineapple to a severe-looking king. Growing a pineapple, the most exotic of all the fruits, in England was no mean feat, so imagine what he could do with grapes. Indeed his book contains much advice that is still relevant to this day, such as not to plant vines in very fertile soil as this would lead to an overproduction of foliage rather than grapes.

Rose's timing, however, was not good. As John Evelyn and other fine West Country cider makers discovered to their cost, these 9 or 10% alcoholic drinks were appearing at a time in the 17th and 18th centuries when the British were getting a taste for spirits like gin and fortified wines such as port, sherry and Madeira, which would have contained double the amount of alcohol. Even the reds from Bordeaux were usually pumped up for British tastes with strong southern French wines or even brandy. Delicate lower-alcohol drinks like cider were out. Eighteenth-century German historian Baron von Archenholz noted, 'In London they liked everything that is strong and heady.'

Such was the poor reputation of English wine that when Charles Hamilton, the Duke of Abercorn, gave his guests wine from his property at Painhill, which was in Cobham, Surrey, he made sure not to tell them it was English before they tasted it. He planted the vineyard in 1740 and very sensibly employed a French vine grower, David Geneste. The grape varieties used were auvergnat and something called miller – probably pinot meunier. Initially they tried to make a red wine, but the results were a disaster – 'harsh and austere' according to Hamilton. But the white and a pale rosé made from red grapes were much more successful. Hamilton wrote: 'Both of them sparkled and creamed in the glass like Champaign (sic).' The wine sold for 50 guineas a hogshead (300-litre cask) at a time when a barrel of Château Margaux would have cost around 45 guineas. The Champagne comparison is a tantalising glimpse of what might have been had what Hamilton and Geneste learned at Painshill been continued. The vineyard was just a part of the lavish garden on the property, which was one of the first in England designed to look natural rather than in the Classical, formal style that had till then been fashionable. Hamilton imported plants from America and had follies like a ruined abbey built in the grounds. He spent so lavishly that by 1773 he was severely in debt and had to sell the estate. Wine production continued for a number of years after his death in 1786 but by 1814 the vines had been pulled up.

The erratic weather of the era meant you needed deep pockets to make wine in Britain. And pockets didn't come much deeper than those of John Patrick Crichton-Stuart, Marquess of Bute. He lived from 1847 to 1900 and at one point was thought to be the richest man in the world, meaning his vineyard wasn't necessarily run to turn a profit. In 1875, under the supervision of his head gardener Andrew Pettigrew, Bute planted 2,000 vines including gamay, the grape of modern-day Beaujolais, at Castell Coch, a Gothic revival castle near Cardiff.

The satirical magazine *Punch* joked that it would take four men to drink a Welsh wine – two to hold the victim down, and one to pour it down his neck. Nevertheless, the best vintages, like the 1881, were said to taste like still Champagne. In good years they even attempted to make a red wine – the 1893 sold for 60 shillings a dozen, about the same as Chambertin, a smart red Burgundy. But good vintages were rare, with 1882 and '83 both disastrous. Large amounts of sugar were needed every year to boost meagre alcohol levels, and the whole operation sounds like a bit of struggle. Pettigrew reckoned that out of 44 vintages, they only ripened the grapes thoroughly in seven. A journal of the time was 'impressed with the success of the experiment from a climatic point of view, but its costly nature and its questionable pecuniary utility were recognised as being against the adoption of vine-growing in this country'.[4] If I were Bute, I'd have just moved to France and made wine there.

In 1897, Hatch Mansfield,[5] the noted City of London wine

---

4 The *Wine and Spirit Gazette* (*Harper's Weekly*), 16 April 1892
5 Yes, the same firm that planted a vineyard in Kent in 2017 in partnership with Champagne Taittinger

merchant, took on the property aiming to bring it to a wider customer base. Later, managing director Ralph Mansfield commented,[6] 'I do well remember my father telling me of an experiment in selling Welsh wines which was not exactly a success.' Wine production at Castell Coch died out in the First World War, mainly because of the difficulty of obtaining sugar, and the vines were finally uprooted in 1920. The old vineyards are now a council estate.

Pettigrew surmised that the problem wasn't just the average temperature in Britain – there are parts of Europe which grow grapes successfully with a lower average temperature throughout the year – but that in southern England and Wales it rarely got hot enough during the summer. English viticultural expert Stephen Skelton thinks that you need summer temperatures to exceed 30°C on occasion for grapes to achieve full ripeness.[7]

There's no record of any wine being made commercially between the wars, but in the 1940s and '50s the seeds of a wine industry were planted by a few pioneering men and women. The first was Ray Barrington Brock, a former chemist who applied his scientific mind to the problem of which grape varieties would work in the English climate. In photos, with his closely cropped hair and short-sleeved shirts, he looks every inch the modern fifties man. At his home in Surrey he started what he grandly called the Oxted Viticultural Research Station. Here he obtained a variety of vines from across England and the Continent, and planted them to see which would work. His research was invaluable in an industry with no continuous traditions – if, indeed, one could call it an industry; in reality

---

6   *A Tradition of English Wine*, Hugh Barty King
7   *Wine of Great Britain*, Stephen Skelton

there were just a few eccentrics struggling to grow grapes. England in viticultural terms was like a new-world country, only a cold, rainy new-world country, rather than a warm, optimistic one like Australia.

One of the grapes that Brock picked up from Germany was müller-thurgau, which became a staple of English wine. There were countless other varieties he planted that failed to thrive. In the early years the wine wasn't much good either. A vintage was made by Edward Hyams which he described with commendable honesty as 'very poor and yeasty'.

Yet even if the results were underwhelming, Brock's and his peers' work was a great help to those who would come later. Every English vineyard planted since owes something to Brock's trial-and-error approach to ripening grapes. Hyams was perhaps even more important because through his books and media work he disseminated information about viticulture and winemaking. Part of the problem in the past was that important facts would be learned but then forgotten. Hyams made sure that this did not happen and helped ensure that winemakers could learn from each other's mistakes.

Despite their importance in the development of English wine, neither Brock nor Hyams made wine commercially. The pioneer here was – deep breath – Major-General Sir Guy Salisbury-Jones, who in 1952 planted a vineyard at Hambledon in Hampshire, the village famed as the 'cradle of cricket'. Born in 1896, Sir Guy had served on the Western Front in the First World War and picked up a love of France including French wine. Later, after the Second World War, he served as military attaché in Paris from 1946 until 1949. Returning home, he was determined to make good-quality dry wine in England.

The chalky soil at Hambledon is very similar to parts of

Champagne and is today planted with chardonnay, pinot noir and pinot meunier. But inspired by Brock's work, Sir Guy planted müller-thurgau and seyval blanc, a French hybrid.[8] If you want to see what the early days were like, there's a wonderful Pathé film on the Hambledon website of a harvest, probably in the early 1960s, where all the men are wearing ties, a sartorial standard to which I wish modern winemakers would return. But don't let the 'jolly day out' feel of the film fool you. Hambledon was a thoroughly professional affair. Winemaking was in the hands of Bill Carcary, who lived until 2021 and continued to help out until just before his death, in conjunction with Austrian consultant Anton Massel, who introduced more consistent winemaking techniques from Germany. Massel would later go on to found the IWSC (International Wine and Spirits Competition).

Sir Guy's example inspired others to plant vineyards and attempt to make wine. In 1958, Margaret Gore-Brown planted vines at Beaulieu House in Hampshire (she later gave her name to the trophy for English wine of the year at the UKVA, now Wine GB, annual awards). It was a revival of winemaking on the property that had previously been carried out by Cistercian monks in the 13th century. Indeed there's a great story of how Beaulieu Vineyard in California tried to stop Gore-Brown selling wine under the name of her house before she informed them that they had been making wine at Beaulieu since before America existed. The matter was quietly dropped.

By 1967, there were enough growers to form an English Vineyards Association. The first chairman was Jack Ward of the Merrydown Wine Company based in Horan in East Sussex.

---

8   See 'Grape Expectations', chapter 12

Despite its name, Merrydown was a specialist in cider and country wine (wine made from fruit other than grapes) but in 1954 Ward had been inspired by Brock to plant some grapevines. English wine was finally building up a bit of momentum. There were even competitions, including at Christie's, where home-grown products were tried blind against German and French ones and not found wanting. Yet some people still had trouble with the very idea of English wine. Sir Guy told a story of trying to bring some of his wares into France for export and being stopped at Orly Airport: 'The wine seems to have aroused the suspicions of the French douane [customs officials] who could not believe that England produced wine.' Steven Spurrier had a similar problem when he tried to organise the delivery of an English wine to be served to President Georges Pompidou and Queen Elizabeth II at a dinner at the British Embassy in Paris. He was curtly informed by French customs that 'English wine does not exist.'

The weather still wasn't cooperating either. In *A Tradition of English Wine*, Hugh Barty-King describes 1954 to 1958 as a 'succession of bad years'. Only the very best sites could ripen grapes sufficiently to make a drinkable wine. By the sixties, Brock had realised that his vineyard in Surrey was perhaps not so well sited at 450 feet (137 metres) above sea level. The extra altitude meant that the grapes ripened late. He wrote 'It used to be said that every hundred feet of height represented three days' lateness of ripening of ordinary fruit.' Like Brock, other English growers were beginning to work out what was needed from a vineyard site. Situation was absolutely critical, in a way that it wouldn't be in warmer climates like Burgundy or the Loire. In France it might be the difference between good and great wine; in England it was the difference between something drinkable and something that would be better off turned into vinegar. Ideally the vineyard should be close to sea level, on a south-facing slope to catch the evening rays of sunshine. Shelter from high winds was important, so not too close to the sea, but proximity to large bodies of water was helpful in preventing the temperature dropping too low. Good drainage was important – vines don't like wet feet – and frost traps such as ditches or wet, low-lying fields were to be avoided.

What were these early wines like? English wine writer Oz Clarke says, 'I have not found the slightest evidence that anything Brock made was drinkable.' He describes the early Alfriston English Wine Festivals he attended in the 1970s as 'sodden affairs with sodden people giving you sodden wine'. He opened an old English wine recently, 1976 Chilsdown from West Sussex, and deemed it 'unbearable to put it in my mouth'. He added, though, that there were rays of sunshine like Biddenden and Spots Farm in Kent, and Breaky Bottom in East Sussex.

In the late seventies and eighties, a second wave of producers was to arrive, who were more commercially minded, and better trained and equipped. Using techniques imported from Germany by men like Anton Massel, they were able to temper the high acidity of England's grapes with a little sugar, making them a lot more palatable. It was in the eighties that English wine would experience its first boom.

CHAPTER 2

The bloody awful weather years

'The industry was the preserve of crusty old major-generals, wealthy farmers and a scattering of military retirees'

Stephen Skelton

Visiting Peter Hall at Breaky Bottom in Sussex is a pilgrimage for any devotee of English wine. Though only ten miles from Brighton, it feels like you could be in a remote Welsh valley. It's perhaps apt, then, that Hall started out as a shepherd before marrying the local farmer's daughter and becoming, with his new wife, tenants on a six-acre small holding.

Born in 1943 and brought up in Notting Hill, 'when it was still rough', Hall studied agriculture in Newcastle. He had no experience of viticulture or winemaking but he grew up in a wine-drinking household – unusual for the time. His mother was French and her father used to own a famous restaurant in

Soho called Au Petit Savoyard which was popular with showbiz types, lords and gangsters. In 1974, Hall was inspired by a vine grower on the Isle of Wight called Nick Poulter (the author of two books – *Growing Grapes* and *Wine from Your Vines*) and planted müller-thurgau and seyval blanc vines on the slopes surrounding his Sussex cottage.

He's been there ever since, making wines that have become legendary in the English wine canon. It helps that Hall looks somewhat mythological himself, like the ancient mariner with his striped sweater, sailor's cap and the constant smoking of roll-up cigarettes. I'd heard that he was a whisky fan so brought him a bottle when I visited, though he told me that at his age he doesn't drink very much these days, and we stuck to tea. Probably wise as I was driving and there's a two-mile unmade track in and out of Breaky Bottom – which is named after bracken rather than what the drive did to the underside of my old Mercedes.

Breaky Bottom has established a reputation that resonates beyond the small world of English wine. Hall himself seems torn between an urge to cut himself off in his little valley and his delight that the outside world seems so interested in what he is up to. Just before I visited, he'd been to a reception at Clarence House hosted by Camilla Parker-Bowles (now, of course, the Queen Consort, though this was before the death of Queen Elizabeth II) and was fresh from being interviewed by the German wine press. He proudly told me how the wine buyer from Sketch, a fashionable London restaurant, was demanding: 'I want Breaky Bottom,' while Christina, his second wife, was reminding him that he needed to label some bottles for the wine merchant Corney & Barrow as the buyer was coming to collect them in person the next day. And it's not just the wines that are in demand; Hall has had to fend off potential buyers for his farm. Shaking his head, he told me[9] about one suitor from the Low Countries who had offered him ten times what he paid for the property when he bought it outright in 1994 for £100,000.

It's a far cry from scratching a living in the seventies, which Hall recalls was how English winemaking was back in those days. In 1976, his first harvest, he had some beautifully ripe grapes but no winery, so he took the crop to one of the largest commercial operators at the time, Lamberhurst in Kent which was owned by Kenneth McAlpine,[10] a former racing driver and scion of the construction family. The winemaker was a German called Karl-Heinz Johner who had trained at the country's elite wine school at Geisenheim. Hall wanted a dry, French-style wine rather than the sweet Germanic wines that most English producers made. Eventually the time came to try what Johner had made and immediately Hall could tell that the winemaker had 'completely goofed it up'. Johner then admitted that, despite his education, this was his first vintage. 'Just my luck to make wine for someone who knows about wine,' he said, as Hall recalled in a spot-on German accent. A morning with Hall is like spending time with Rory Bremner.

The problem was that Johner had applied formulaic, cold-climate winemaking practice to Hall's grapes. This involved adding water to bring down the acidity and then adding sugar before fermentation to make up for the dilution. But because Hall's grapes were not the searingly acidic grapes that were

---

9   He also told me, with a gleam in his eye, that the myth of putting a spoon in the top of a bottle of Champagne to keep the fizz in was put about by the 'powerful spoon lobby'

10 McAlpine died on the 8th of April 2023 at the age of 102

normal in England, the subsequent wine was lacking in acidity. To make up for this Johner added citric acid and some unfermented grape juice at the end of the process to sweeten it – a process routinely done to cover up for unripe grapes, which Hall's weren't. He then topped it off with a dose of sulphur to stop the wine refermenting in the bottle. It tasted awful. Hall submitted a claim for the value of the wine that had been ruined and, after more than a year of letters and phone calls, eventually got the money out of McAlpine. The irony is that Lamberhurst was probably the best equipped winery in England at the time. If it was turning out bad wine, then there's every chance that everybody else's was worse.

By the time of the next vintage, Hall had established his own winery. He showed me some photos: it looked incredibly primitive, making wine in what looked like large buckets. 'There was no money for proper tanks or pumps,' Hall outlined. Thirty years later, Johner, now a successful winemaker, came to visit Hall, who says there were no hard feelings.

Incompetent German winemakers weren't the only problem that Hall had to deal with. For five straight years, water built up on a neighbour's field before pouring down the hill and flooding Hall's cottage to the extent that he and his wife were forced to move into a caravan. Fortunately, the vines planted on the steep hillsides were largely untouched by the deluge. Then there was the plague of pheasants that ate all his grapes one year. There was something biblical about the trials that Hall faced and eventually overcame, but I get the impression that he relished a fight.

The hardest thing of all was what Hall describes as the BAW years – bloody awful weather. In the seventies and eighties, winemakers would be lucky to get two good vintages a decade, with six passable and two that would be a complete write-off. Most of the time Hall was making what he describes as 'tolerably acceptable still wines' – though they were considered by Oz Clarke and other wine writers to be England's finest.

One of Hall's great friends is the former winemaker Stephen Skelton. These days Skelton is Mr English Wine, a Master of Wine who has written many books on the subject and now works as a consultant for those looking to plant vines. Skelton established a vineyard at Tenterden in Kent that would eventually morph into the present-day behemoth Chapel Down. Unlike almost every other English winemaker, Skelton had had proper training at Geisenheim. He first planted vines in England in Easter 1977 and wasn't impressed with the competition. In fact, he said at the time that he had never had an English wine that he liked. Skelton, like Hall, makes marvellous company because he has strong views on everything, which he has no hesitation in expressing. There's something of the Professor Yaffle, the learned irascible woodpecker from *Bagpuss*,[11] about Skelton, tutting and laughing at the incompetence and foolishness of other English winemakers – and, no doubt, other wine writers.

Skelton's first vintage was in 1979, of which he produced 8,000 bottles. But it was the next year that put him on the map. He made a seyval blanc with a little bit of sweetness in it and some carbon dioxide left over from fermentation still bubbling through the wine – a 'spritz' rather like you get in some young riesling kabinetts or a vinho verde from Portugal. Much to Skelton's surprise, he got a call out of the blue from Tony

---

11 A whimsical British children's programme from the 1970s made by Oliver Postgate

Laithwaite, who ran the UK's largest mail order wine merchant, asking to buy a few cases. This wasn't Laithwaite's first foray into English wine – the company had listed a 1974 Adgestone from the Isle of Wight,[12] a vineyard that is still in business. But this time Laithwaite had been tipped off by Hugh Johnson, chairman of the Sunday Times Wine Club, a partnership between the newspaper and Laithwaites, who had tried the wine at a competition.

A few weeks later, Skelton had another call, this time from Colin Gillespie from UKVA (United Kingdom Viticultural Association) informing him that he had won the Gore-Browne trophy for English wine of the year. This was the competition that Johnson had been involved with, and he'd tipped off Laithwaite who'd cannily managed to buy a few cases before the results were known and Skelton put the price up. Gillespsie added, however, that some on the judging panel thought that Skelton had cheated 'by having trained in viticulture and oenology [winemaking]'. There was something not quite cricket about knowing what you were doing, which went against the amateur ethos of English wine at the time. As wine journalist Jancis Robinson, who then wrote a profile on Skelton for the *Sunday Times*, said on her website: 'For several decades, vine-growing was typically a retirement occupation for the well-heeled with a paddock to spare . . .' Or as Skelton himself put it:[13] 'The industry . . . was the preserve of crusty old major-generals, wealthy farmers and a scattering of military retirees.'

---

12 In some ways, as the home of Adgestone and the original home of Chapel Down, the island is one of the cradles of English wine but it's not much talked about. This is likely to change soon as the owners of Coates & Seely, the award-winning sparkling wine from Hampshire, have acquired 23 hectares there and will be planting in 2023

13 Academieduvinlibrary.com

But the blazer and regimental tie brigade of the fifties and sixties was giving way to people who were determined not only to make wine worth drinking, but to make money out of it too. One such was Christopher 'Kit' Lindlar who is now a Roman Catholic priest in Norfolk. I spoke to him on Zoom, Lindlar sitting in front of the church altar in his dog collar. In the past, all you needed to be a wine producer was 'a major's pension and a double-barrelled surname', he told me. Instead, Lindlar went to study in Germany, though he referred to it as a technical high school course rather than a degree like at Geisenheim. 'I went and lived in Germany for a couple of years, got an interest in wine, and thought I might make a career out of it. I found the technical side much more interesting than the marketing side,' he explained. Lindlar came back to England during the searingly hot summer of 1976. His experience in Germany was enough to land him a job making wine for Merrydown, the cider company set up by Ian Howie and Jack Ward, founder of the UKVA, and which had a contract wine arm that apparently never made much money. 'They were perennially broke,' Lindlar said, 'but it was a very small part of the business and dear to the heart of Ian and Jack.'

Merrydown's fortunes were turned around when it won the contract to make cider for Marks & Spencer. Wisely, Howie and Ward decided to hive off the wine business to Lindlar who was, in any case, planning to set up on his own and poach Merrydown's customers, 'so it was very good timing'. Initially Lindlar used Biddenden, a winery set up in 1969 in Kent, to bottle his vintages, but he later built his own winery at High Weald, also in Kent. He'd get grapes from all over the south-east. From these he made Germanic style off-dry wines which would weigh in at about 10.5% ABV (alcohol by volume) with

31

about 5–10 grams of sugar added in the form of grape juice. 'In a good year with good alcohol you could get away with bone dry but otherwise a touch of sweetness brought the flavour out,' he said. 'In those days all the technology, grape varieties and winemaking equipment came from Germany.' France didn't make tanks and other equipment in small enough sizes for the nascent English wine industry.

Lindlar estimated that in the mid-eighties, around 70–80 per cent of wines made in England would have been made either by himself, Stephen Skelton at Tenterden or by our old friend Karl-Heinz Johner at Lamberhurst. At the time, Lindlar organised a comprehensive blind tasting of English wines at the Royal Commonwealth Society where the scores were analysed by the University of Kent. The results revealed the most important factor in determining a wine's quality wasn't where the grapes were grown, but who had made it. Nevertheless it was, of course, important that the grapes were of good quality. 'You needed really good management and weather,' said Lindlar. 'At High Weald I had a system where if people wanted to send grapes [to be made into wine] they had to sign an agreement that I would visit three or four times a year and give advice on vineyard management and improving grape quantities.' Even so, in some cases it was a case of 'making a silk purse out of a sow's ear', he admitted, with grapes often planted in unsuitable locations. Lindlar recalls some of the people he was dealing with: 'Some bloke made a stack of money in the City and bought a house in the country with plenty of land for his daughters' horses. Then the daughters discover boys, so what to do with the land? I know, I'll plant a vineyard.'

But the sort of people moving into wine was changing. One of the new breed was an ex-RAF pilot called Bob Lindo. Lindo's aviation career had ended abruptly with a mid-air collision and a broken spine. He was lucky to have survived. At 38 he wanted to start all over again, and in 1982 the family bought an 82-acre farm near Bodmin in Cornwall. The original plan was to keep sheep and other livestock, but the Lindos noticed that the south-facing slopes were ideal for growing grapes. 'The place inspired us,' Lindo recalls. 'It was a very exciting time. The first wave of English winemakers were millionaires who didn't know what to do with their money. I started at the very beginning with nothing.' People like Lindo brought a hard-nosed commercial outlook to the industry. Wine was their income, and their ventures had to be profitable. 'Now it's seen as a business,' he adds.

If you didn't want to go abroad to study, the only way to learn about wine was from other English winemakers. Lindo was lucky in that Gillian Pearkes wasn't so far from him in Devon. Pearkes was a pioneering winemaker and teacher who planted Yearlstone Vineyard in 1976 and wrote a number of books on growing grapes in Britain.[14] Lindo paid £25 to do a course with Pearkes to learn the rudiments of viticulture and winemaking, and bought some vine cuttings from Derek Pritchard who is still in the vineyard supply business today. Lindo describes Pritchard and Pearkes as his 'two heroes of the time. I learned so much just from sitting in a lorry next to Pritchard and listening to his wisdom'. He also read every book he could get his hands on.

Lindo wasn't entirely new to winemaking. In the seventies and early eighties, economic conditions were hard and a lot of people turned to home winemaking, rather like Tom and

---

14 See bibliography

Barbara in *The Good Life*. 'Everybody had done a bit of wine-making in those days because nobody had any money,' Lindo says. Lindo and his wife Annie planted their first commercial vines in 1989. 'We paid for professional advice and planted 8,500 vines over five acres,' says Lindo. Even then, there was an element of amateurism to the venture. 'We just dug 8,500 holes and put them in.' They did everything themselves. 'When we were here there was nothing, not even a road, just a dilapidated house. We built a winery by hand for £15,000 – two people in two years.' The winery was named Camel Valley.

Lindo told me how his first sparkling wines were all made by hand using an electric drill to mix in the sugar and the yeast, while after disgorging (removing the yeast sediment) he used a sheep syringe to add the dosage (the sugar solution to sweeten the wine before corking, see glossary for fuller explanation). 'I had a proper talent for it and liked nothing better than getting my hands dirty changing a filter,' he says. Despite the DIY ethos, Lindo knew there was no point growing good grapes only to then mistreat them in the winery, so he did eventually purchase high-quality equipment. Whereas most of the English wine industry was using fibreglass tanks, Lindo invested in stainless steel, bought direct from wine trade fairs he attended in Bordeaux.

One of the first people who realised that Lindo was doing something special was local restaurateur and TV personality Rick Stein. Stein's restaurants in Padstow have remained some of Camel Valley's best customers and Lindo featured on the BBC TV series *Rick Stein's Food Heroes* in 2004. Like Hall, Lindo has become feted within his industry, his wines served to royalty, and he's won multiple trophies culminating in a Lifetime Achievement Award at the International Wine Challenge in 2018. It would also be fair to say that Lindo has a flair for publicity. Oz Clarke tells the story[15] of Lindo camping outside Gatwick airport to protest about a 2012 London Olympics wine promotion that featured Champagne but no English wine. The press had a field day. He also campaigned successfully to the EU to have a designated PDO for bacchus from his Darnibole vineyard in Cornwall. The winery is still very much a family business, with Lindo's son Sam now making the wine. One of the wines produced is called Annie's Vineyard, after Sam's mother who looks after her own 5,000-vine plot.

Though it was modest compared to what would come later, the mid-eighties marked the start of the first English wine boom. 'There was an enormous expansion with more vineyards, though still on a relatively small scale,' says Lindlar. Press interest increased and mainstream shops and supermarkets began stocking English wine. The product was getting better too, something that Lindlar puts down to more professionalism, but also to the vines being older (better-established vines usually produce higher-quality wine). The strong economy, at least in the burgeoning financial sector under then chancellor Nigel Lawson, helped too. According to Lindlar, 'A lot of City types bought country houses and planted vineyards.'

The biggest of them all was Denbies in Surrey. The vineyard, established in 1986, was the brainchild of Adrian White, who made his fortune with an engineering company, Biwater, and was vast – 247 acres at a time when a 20-acre English vineyard would have been considered notable. Oz Clarke remembers 'a huge fanfare' about the arrival of this new player in English wine, in many ways a harbinger of things to come. But Clarke[16]

---

15 *English Wine*
16 *English Wine*

wasn't impressed by the results: 'The wine just wasn't any good,' he recalls. Lindlar worked one vintage at Denbies and says that the problem was that White wasn't a wine drinker: 'He was very ambitious for it as a project but it wasn't entirely clear what the ambition was.' According to Lindlar, White 'wanted results now and wasn't prepared to play the long game and let the winemaker get used to it'. As a result, White got through a succession of winemakers. Almost among them was the now-celebrated Owen Elias (who has served as winemaker at Chapel Down, Balfour, Nutbourne and Kingscote, among others) who recalls nearly applying for a job at Denbies when his role at Chapel Down was looking uncertain but deciding against it as 'the wines were so disgusting. I didn't want to make wine like that.'

White's timing also wasn't ideal. Denbies was planted with German varieties at a time when British drinkers were developing a taste for Australian chardonnay, New Zealand sauvignon blanc and Chilean merlot rather than the off-dry German wines which had been so big in the sixties and seventies. Grape varieties were becoming brand names but England had all the wrong brands. According to Sam Linter, winemaker and managing director at Bolney in Sussex, these German vines weren't even very well suited to the damp English climate, and were particularly susceptible to rot and mildew. This meant that the grapes were often picked too early, when they weren't ripe. Or as Stephen Skelton wrote:[17] 'Wrong varieties, wrong sites, poor winemaking, poor quality and lastly and more importantly, bad marketing.'

By the late eighties, the mini English wine boom had turned to bust. Producers had far too much stock that wasn't selling. The largest of them, Lamberhurst in Kent, reduced its prices

---

17 *The Wines of Great Britain*

dramatically, forcing others to do the same. Many producers went out of business and vineyards were pulled up. Stefano Cuomo, whose family runs Macknade, an upmarket food hall in Kent, describes many producers of the time as 'growers with no understanding of marketing'. He went on to say that there was a 'farm-shop mentality' with wines packaged like jam. These inconsistent, weedy, strangely-packaged wines couldn't compete with the likes of Lindeman's Bin 65 Chardonnay that were coming out of Australia.

It's fitting, therefore, that it was an Australian, John Worontschak, who would eventually turn around the fortunes of Denbies when he arrived in the 2000s. It would be the start of a small invasion of Antipodeans or those trained there. David Cowderoy, whose father Norman planted Rock Lodge vineyard in West Sussex in 1961, trained at Roseworthy in Adelaide. According to Owen Elias, who worked as his assistant winemaker in the nineties and was paid £2.50 an hour for his troubles, it was 'only him and Worontschak who had the faintest idea; everyone else had come through Geisenheim'. Cowderoy introduced techniques that were common in Australia at the time, such as adding enzymes, tannins and oak chips to fill out the flavour profile of wines. He had high-tech equipment like a Diem tank press – 'beautiful juice but a nightmare to use,' according to Elias – and pioneered temperature-controlled fermentation in stainless steel at a time when most producers were still using fibreglass tanks (today, these sort of tanks are used to store cider; you can see them outside Biddenden).

Elias describes Cowderoy's techniques as 'tricks of the trade', and simply 'really good winemaking'. Not everyone, however, was impressed. Lindlar says: 'In those days, the problem was these young Australian winemakers churned out by Roseworthy who were trying to turn the rest of the world into South Australia. [The wines were] great to drink in the morning, less interesting after. That's the trouble with Australian wine – it's overpriced and over here.'

But as confidence increased, so winemakers in England weren't looking to Australia, or indeed to Germany. They were inspired by France. And spearheading the French invasion of English wine was a half-French, half-English winemaker from Bordeaux called Chris Foss who came back to Britain in the mid-eighties. Foss had been making wine in Bordeaux both at his family's property and at larger operations like Chateau d'Yquem in Sauternes, one of France's most prestigious producers. He had a shock during his first English vintage, when, after a four-week summer holiday in France, he returned to find the weeds around the vines three feet high. Not something that would happen in the baking heat of a Bordeaux summer and a sign that making wine in England was going to be very different.

It wasn't the only shock he got. He recalls buying a case of wine from the English Wine Centre in Alfriston in East Sussex. 'Over half of them were not commercial. They were out of balance, had wine faults, the labelling was bizarre and the prices were high.' It was all a long way from d' Yquem.

Foss never intended to set up a wine school or indeed to make wine in England. When he came back to the UK from France in 1984 it was because his family's vineyard wasn't making money and his wife wasn't happy in France. He sold the farm in Bordeaux for 'the price of a semi-detached house in Burgess Hill', trained to be a teacher but ended up looking for a job in wine. Such was the dominance of German expertise in English winemaking at the time that when he turned up at vineyards, he would constantly be asked 'Why aren't you

German?' Eventually he was offered a job at St George's vineyard in Horam, 'but they couldn't pay me enough' so he was told to try Plumpton, a small agricultural college just outside Brighton which opened to students in 1927 and was apparently considering launching a winemaking course. Foss went along, told them he knew how to make wine and had a teaching qualification, and they offered him the job on the spot. It was a case of, 'here's a desk, just get on with it,' Foss says. The local council gave the college some money to buy a tractor and a sprayer but, 'for the first three years I was a one-man band'. The first intake of students was in 1988. Equipment was rudimentary, the winery was in a chicken shed with some glass demi-johns in which to make the wine. The largest tank was 100 litres. There were only two rows of vines to cultivate, so in the first year the college bought grapes from Barkham Manor. 'Little by little we planted [more] vines,' says Foss. Then, 'Seven or eight years later, St. George's vineyard came up for rent and we took it.' Gradually they also took over other local vineyards like Ditchling and eventually David Cowderoy's Rock Lodge.

One of Foss's first pupils was Owen Elias, who was looking to change careers following a disastrous experience in the music industry.[18] He did a viticulture and oenology course at Plumpton, which involved him coming down from the Lake District for short periods. At one point he was living in a tent next to the college. He described the set-up at the time as 'ridiculous – a garage with demi-johns'. Nevertheless, he is full of praise for the quality of the education he received, which expanded as Foss began to offer more in-depth courses.

'We had more and more people from the trade coming

---

18 See 'Not Going Tits Up', chapter 5

through and I was asked whether I wanted to do a proper full-time course in conjunction with the University of Brighton,' Foss says. Initially, most of the students were people who owned their own land and wanted to learn how to make wine, but gradually, younger students came in who were looking for a career. In those early days most students would end up working abroad in New Zealand or France – I met my first Plumpton alumni at a vineyard in Faugères in the Languedoc – but it's a mark of how much the domestic industry has grown that most now stay in England.

Foss, as much as anyone, is responsible for introducing consistency and professionalism to English winemaking. Today, when you buy a bottle of standard English wine and it's tasty, you have Foss to thank. What started as a one-man band for English wine has become the equivalent of Roseworthy in Australia, Montpellier in France or UC Davis in California.

It's hard to imagine those early days in the chicken shed when you visit Plumpton now. A snazzy new building was opened by Jancis Robinson in 2007 followed by a £500,000 extension in 2014 including the Rathfinny Research Centre, paid for by the winery of the same name which at the time had yet to release a wine, and a Jack Ward Laboratory named in honour of the English wine pioneer from Merrydown. The college has a commercial arm that regularly wins awards for the quality of its wines, all of which are made by the pupils. They get a very broad education, with trips to Georgia to learn about making wine in qvevri (clay jars that are all the rage these days). The college offers everything from a BSc in viticulture and oenology to degrees in wine business and short courses that restaurants and retailers can send their staff on.

Following in Elias' footsteps, today's head winemaker and

assistant winemakers at Chapel Down, England's largest producer, are both Plumpton-trained. Indeed there probably isn't a winery in the country that doesn't have Plumpton alumni on the pay-roll. Liam Idzikowski from Danbury Ridge in Essex describes it as 'hugely important, practical and affordable', and also notes that Plumpton qualifications help keep the talent at home. 'If I'd studied at UC Davis or Roseworthy, I'd have made wine abroad.' In addition, the college provides a networking opportunity where winemakers can swap information and help each other. The industry has become more collaborative thanks to work at Plumpton. As John Atkinson, also at Danbury Ridge, says, 'if we didn't have Plumpton it would be all closed doors.' He compares a Plumpton qualification favourably with one issued by the Institute of Masters of Wine; he is himself a Master of Wine, which is an expensive and exclusive membership organisation, whose notoriously challenging qualification 'offers lifestyle opportunities at a price', as he puts it.

Foss happily admitted to me that he knew nothing about English wine when he started. 'I was working with varieties I had never heard of and then trying to get them to ripen.' He tried to grow chardonnay but it was a disaster. 'We produced one demijohn with acidity of 25g per litre, and it was horrible.' To put that into context, 25g per litre is more than double the level you'd expect to find in an English sparkling wine today. It would actually hurt to drink a wine that acidic. One of the most important techniques taught at Plumpton was, therefore, de-acidification. This was done by adding chemicals like calcium carbonate, essentially chalk,[19] to the wine – one reason why old-school English wines often taste chalky. Alcohol levels would

---

19 It's helpful to think of antacids like Gaviscon here

need boosting with sugar: 'The spend on sugar was huge,' says Foss. Fungi like powdery and downy mildew were also problems though botrytis, another fungus, wasn't an issue simply because the grapes didn't get sufficiently ripe.

Whereas the 1970s were, in Peter Hall's memorable phrase, the BAW years, by the late eighties and nineties, something was changing. Previously, you'll recall, Stephen Skelton[20] 'didn't see a single day when the temperature climbed above 29°C'. But Skelton noticed the weather warming up in 1989 and 1990, with 1994 the year things really took off, right up to the scorching 2003 vintage. After this there was a rash of plantings as people realised that England was becoming a place where grapes could be ripened nearly every year.

'The factor that has most influenced viticulture in Britain in the last three decades is that of climate change,' Skelton says. Foss agrees: 'I didn't realise that climate change would play such a big part in what we do. It has made things completely different. We are now a cool climate, not a cold climate. We have the same climate that Champagne had in the 20th century.' Foss now thinks that southern England is too warm for müller-thurgau, as the grapes don't have enough acidity – something that was scarcely conceivable when he started.

The changing climate allowed English producers to move away from the decidedly unsexy German varieties, and on to French grapes that people had heard of. Professor Steve Dorling from the University of East Anglia's School of Environmental Sciences wrote of how things changed between 2004 and 2021: 'Over that period the warming climate supported much more reliable yield and quality of the pinot noir and chardonnay

---

20 *Wines of Great Britain*

grape varieties.' These are, of course, the two main grapes that make Champagne.

One of the first things Foss did when he launched his course was to organise a trip to Champagne, the nearest fine wine region to England and the closest in terms of geology and climate. Apparently the Champenois, 'thought we were very funny', he recalls. Little could they have guessed what was coming. In that first intake of students at Plumpton were two people with big ideas: Sandy Moss, who founded Nyetimber with her husband Stuart, and Mike Roberts, who would go on to found Ridgeview. The three of them were to change the face of English wine . . .

CHAPTER 3

Ambition and money

'We couldn't figure out why we couldn't, so that meant we could'

Stuart Moss

Reading histories of English wine can be a tantalising, frustrating business. In hindsight, the Champagne model was always the way to go. Sparkling wine requires grapes with high acidity, and the process[21] used to make bubbles softens that acidity and makes it not only palatable but positively stimulating. England's cool climate means that grapes will often be too acidic to make palatable table wines but such grapes are the perfect raw material for sparkling wines. As the grapes become riper, the acidity drops, which is why hot climates like the south of France don't produce world-class sparkling wines.

---

21 We'll look at this in closer detail in 'Fizz Wars', chapter 8

In the rare cases where winemaking flourished in Britain in the 19th and 20th centuries, there was often a Champagne comparison. At Painshill in Surrey in the 18th century there's even a mention of the wine foaming like Champagne. This was at a time when cider makers and brewers were learning how to harness bubbles in bottles. Why couldn't people apply that to home-grown grapes?

When English wine was revived after the Second World War, this is just what some producers did. Raymond Brock produced a sparkling wine in 1959 which was compared favourably to Moët & Chandon by one possibly over-enthusiastic Master of Wine. Pilton Manor in Somerset produced bottle-fermented wines in the sixties but it was in the eighties that David Carr Taylor began to make such wines in commercial quantities at his vineyard near Hastings, winning awards and even exporting some. Another winemaker, Owen Elias, told me about making a sparkling wine at David Cowderoy's vineyard Rock Lodge in West Sussex which won the IWSC English wine trophy in 1991. The owners wanted to call it Imperial but that is a Moët & Chandon brand and the mighty Champagne house objected so they called it Impresario instead.

Although it had been noted by English wine pioneers like Gillian Pearkes and Major Sir Guy Salisbury-Jones that southern England had locations with similar soil and climate to Champagne, it wasn't easy to ripen the traditional French varieties here, so these early sparklers were made with Germanic grapes like reichensteiner and huxelrebe. By the 1980s, however, growers were beginning to find the right sites, with shelter and plenty of sunlight, where in good years they could ripen more exotic varieties. At Surrenden near Ashford in Kent, they have been growing pinot noir, chardonnay and pinot meunier

46

since 1984. At New Hall in Essex, they have pinot noir vines dating back to the 1970s. Stephen Skelton writes[22] of a sparkling wine made in 1986 by Karl-Heinz Johner from pinot noir and chardonnay.

So in 1988, when an American couple, Sandy and Stuart Moss, planted grapes at a medieval manor house called Nyetimber in West Sussex with the aim of making sparkling wine, it might have seemed that they were not doing anything particularly revolutionary. But the couple brought with them two things that the industry had been sorely lacking: ambition and the money to back it up.

Stuart Moss was from a Chicago family and had made a fortune producing dental equipment. He met his wife Sandy (née Puetz) when she was working as a receptionist for his company. They were married in 1971, and in 1982 he sold the dental firm and the couple decided to move to England. By this time, Sandy had a business selling antique English furniture. The couple had no experience of winemaking or growing grapes but on a trip to a vineyard in Suffolk the idea came to them to make wine. They began touring England's vineyards but were not impressed with the quality of the wine they encountered.

Shortly before his death in June 2022, I was fortunate to talk to Moss down the line from his home in Santa Barbara, California. He told me: 'People who were involved in English wine were getting expertise from Germany. It's cold in England and cold in Germany so they planted the same varieties to make an off-dry Liebfraumilch style which was fine for a picnic but was never going to go anywhere. They were banging their heads against the wall. We decided to aim right for Champagne, so it

---

22 *Vineyards of Great Britain*

was do or die.' He told me a story about an official from the Ministry of Agriculture who said that they would be better off planting apples. Moss snapped back: 'We didn't move 4,000 miles to grow apples!' He continued: 'It became more and more evident that sparkling wine was the way to go, and the more people said we couldn't do it, the more determined we became.' Not everyone was unsupportive. Author, winemaker and teacher Gillian Pearkes told them, 'Find the right site and you will grow chardonnay in England.'

It took the Mosses five years to find the right site. In 1985 they were interested in acquiring England's original commercial vineyard, Hambledon in Hampshire, but Stephen Skelton remembers that 'they lost out to another bidder, John Patterson, who owned it until 1994.'[23] Then they struck gold. According to Sandy Moss, 'We found the farm by accident. We were looking in the area and we knew that across Gay Street [a hamlet in West Sussex] there was an apple orchard, which meant the same microclimate you need to grow grapes.' The place in question was Nyetimber, a medieval timbered manor house that once belonged to Henry VIII. It was mentioned in the Domesday Book, according to the Mosses, as somewhere that had vines, and sat on 90 acres of greensand soil. They sent soil samples to be analysed in Champagne and the results came back that it would be a good place for viticulture. The nearby Brinsbury agricultural college supplied them with twenty years of meteorological data, which told them it should be possible to ripen the grapes – at least in most years. Such research is routine for prospective growers today but at the time this attention to detail was unusual. Many people just planted in a spare field

23 *Wines of Great Britain*

and hoped for the best. Stuart Moss recalled people planting chardonnay on north-facing slopes with no trellising to keep the vines off the ground, and being surprised when they couldn't ripen it. He continued: 'We had the advantage of knowing absolutely nothing so we had to start from scratch, reading everything and figuring it out. If we couldn't figure out why we couldn't [do something], that meant we could.'

The Mosses acquired Nyetimber in 1986 and the first vines – chardonnay, pinot noir and pinot meunier – went into the ground in 1988. Apparently they flipped a coin to decide who would make wine and who would look after sales and marketing, with the result that Sandy went off to study winemaking at Plumpton College in 1991. Wisely, they also decided that they would need a consultant from Champagne. Through Chris Foss at Plumpton they met Georges Hardy on a tour of Champagne. As the founder of the Station Oeno-technique de Champagne, Hardy knew more than most about making sparkling wine and so the Mosses asked him to work with them. But, as Foss recalls, 'Champagne people didn't speak English in those days,' so Hardy recommended a 'young and dynamic' man called Jean-Manuel Jacquinot who had made sparkling wine in India and spoke English. He was just 25 at the time but as well as working in India and at his family's property in Champagne, had spent time working at Veuve Clicquot. Three decades on, Jacquinot describes to me his experience of English wines back then: 'At that time it was very bad. Still wine was awful, sparkling wine was worse.'

The Mosses had vines, they had money, and they had a French consultant, but to begin with at least, they didn't have a winery. Enter English wine veteran Christopher 'Kit' Lindlar. The Mosses met him at the English Wine Festival at Alfriston and he agreed to make the early vintages for them at his High Weald winery. Long retired from the wine world, he seemed to enjoy sharing his reminiscences of those pioneering days. 'I was sceptical,' he admits. 'It was very daring, it's impossible to underestimate how daring it was. There was no guarantee that it would work.'

For the first three years, the grapes would be sent to Lindlar's winery where they would be made into a base wine. There's some debate as to who actually did the winemaking. Sandy Moss was the public face of Nyetimber and officially the winemaker. But she had only done a short course at Plumpton and Lindlar and Jacquinot were both much more experienced. As Lindlar says, 'It was my neck on the line if it went wrong.' The Mosses then built a very modern winery at Nyetimber which was ready for the 1995 vintage, by when all the winemaking was in-house via a computer-controlled Magnum press – extremely high-tech for the time. Lindlar continued to work there, and elaborated to me on his close relationship with Jacquinot: 'He was very young and I'd been in the business for a while. Two cooks in one kitchen is not a recipe for complete harmony but we got on very well.' Sandy Moss 'acted as umpire between us'. The first official vintage was in 1992, but according to Jacquinot they made a small trial vintage the year before which was successful enough to give them confidence that the Mosses' gamble would work.

Nyetimber's first commercial wine was a 1992 Blanc de Blancs, a wine made from only white grapes, in this case chardonnay. But the Mosses didn't want to release it until it was ready. In 1996, they finally decided that the time was right and held a press tasting at the estate. Nobody came. Still they were undeterred as, according to Jacquinot, 'we knew it was good' having taken samples back to Champagne to taste with French winemakers. Eventually word got out that something remarkable was happening in West Sussex and in 1997 it beat the might of Champagne to win a gold medal in the best sparkling wine category at the IWSC. For the Mosses 'the big one was when Jancis Robinson put a bottle of Nyetimber in with a Champagne tasting'. Robinson served it blind to a group of industry pros who all thought it was top-quality vintage Champagne. Thereafter, the royal household took Nyetimber after hundreds of years of serving only Champagne – 'talk about a total shock,' says Sandy.

The press had a field day. There's nothing the English and American papers like better than the French being taken down a peg or two – especially when wine is involved. It was like the 1976 'Judgement of Paris'[24] all over again. In 1999, *Reader's Digest* – which billed itself at the time as the bestselling magazine in the world – ran a big profile on the Mosses which opened the floodgates. According to Sandy Moss it took months to return all the calls and letters they received after publication of the article.

I was fortunate enough to try the 1992 Nyetimber in 2021 as a guest of Jérôme Moisan, a Frenchman who, somewhat eccentrically since moving to Kent, started collecting English wine. He describes that bottle as every bit as important as Stag's Leap Cabernet Sauvignon 1973, which introduced Californian wine to the world at the Judgment of Paris, or Penfolds Grange 1951, which did a similar thing for Australian wine. It's probably

---

24 A tasting in Paris organised by the late wine merchant Steven Spurrier, where Californian wines beat the best of France in a blind tasting. The event was later made into a terrible film called *Bottle Shock* with Alan Rickman playing Spurrier

Skelton says that Hill had great plans but was knocked off course by a divorce. Yet Jacquinot speaks very warmly about the Hill years: 'It was the same philosophy – to produce the best possible.' Great wines continued to be made at the estate with an Irishman, Dermot Sugrue, taking over the winemaking. But despite awards and rising sales, the estate was still struggling to make money. In a report quoted by Skelton, the estate was working to the assumption that each bottle of wine cost £5 to produce when in fact it was nearer to £9.

After less than four years at the helm, Hill sold up to a man with seriously deep pockets. Eric Heerema is a Dutch shipping magnate and entrepreneur who has since turned the estate into a consistent and widely-admired luxury brand. We'll learn a bit more about him later in the book. As Lindlar puts it: 'The new guy and the team are fantastic. They have achieved what we envisioned in a way that Stuart could not have pulled off.' But that's not to detract from the sheer scale of the Mosses' achievements: 'If it had not worked and we had gone wrong, I don't think what happened subsequently [in the industry] would have happened,' says Lindlar. 'The level of investment was so high they would have been frightened off. This was the crucial moment for English sparkling wine – if it had gone wrong, nobody else would have bothered.'

The Mosses put a rocket under the English wine industry that would show what could be achieved. Or as Stuart Moss said with characteristic modesty: 'We never expected to change the course of history in England. We only aimed to do our own thing.' When Moss died in 2022 while I was researching this book, there was an outpouring of tributes from across the industry and obituaries in *The Times* and the *Daily Telegraph*. I know that Sandy Moss was touched by it all. Brad Greatrix, one half of the current winemaking team at Nyetimber, paid tribute on social media, though there was no official word from Nyetimber, which struck me as a bit odd.

Following his work with Nyetimber, Lindlar would go on to work with another producer who had had the same vision as the Mosses, of making Champagne-style sparkling wine in Sussex. Mike Roberts of Ridgeview studied alongside Sandy Moss at Plumpton, but got his first wine onto the market slightly later. Ridgeview would not only make award-winning wines under its own label but provide a service producing wines for dozens of other brands as well as providing a training ground for the industry. Meanwhile, Jacquinot would go on to work with two of those original pioneers, Peter Hall at Breaky Bottom and Bob Lindo at Camel Valley, as they made their first sparkling wines. England's fizz bubble was ready to explode . . .

CHAPTER 4

Bubbling under

'Oh my God we've won! We turned around and the whole auditorium was giving us a standing ovation'

Simon Roberts

I admit I was rather late to the party with English wine. It was a 2007 Theale Vineyard sampled at a trade tasting in 2013 that made me realise something special was going on. Previously I'd found English sparkling wine a little austere but this one was full, tasting of overripe apples. I described it, clearly trying to channel Oz Clarke, as like 'walking in a cider orchard in autumn'. It was made for Laithwaites wine merchant from surely one of the most bizarre vineyards ever to grow grapes.

It began with a pile of rubble left over from building Laithwaites' HQ at Theale in Berkshire. In 1999, the company's founder Tony Laithwaite, noting that this mound of bricks and concrete faced south, had the idea of covering it in a thin layer of topsoil and planting 800 chardonnay vines, right by the M4 motorway under the Heathrow flight path. The Côte des Blancs it wasn't. And yet for a number of years this tiny vineyard on an industrial estate near Reading produced some of England's best wines. Anne Linder, who has worked for Laithwaites since the 1980s, thinks the idling engines of nearby lorries may have contributed to the particularly warm local micro-climate. Sadly, when Laithwaites moved to new premises, the vineyard was demolished to make way for an Amazon distribution centre, though the vines themselves were rescued and are now thriving at a vineyard in Dorset.

That 2007 Theale that so impressed me came second in a blind tasting of sparkling wines and Champagnes organised by Stephen Skelton in 2013, the so-called Judgement of Parsons Green.[25] The winner was a Wyfold 2009. What did the two wines have in common? Both were made for Laithwaites by Ridgeview in Sussex.

It's somehow fitting that these two illustrious wines were not branded with the name of the winery. As Mardi Roberts, daughter-in-law of Ridgeview founder Mike Roberts, puts it: 'We are not good at showing off. At a strategy day for the business we said that one of our weaknesses was that we're not good at telling people all the amazing things we do.' So while Nyetimber may have got the column inches, far more people were drinking wines made by Ridgeview – even if they didn't know it. Indeed the current owner of Nyetimber, Eric Heerema, told me that the bottle that convinced him that English wine had a future was a 2003 Ridgeview.

---

25 A play on the 1976 Judgement of Paris wine tasting organised by Steven Spurrier

The Ridgeview story runs almost parallel to that of Nyetimber. The company's founder Mike Roberts studied alongside Sandy Moss at Plumpton College in 1991 and the pair's first vintages were both made by Kit Lindlar. Roberts ran an IT company that he began as a business from his kitchen and went on to become the biggest IBM dealer in Britain according to his son Simon. To encourage his staff, Roberts ran something called the '100% club' where the best performing employees were taken on trips to Champagne. It was on one of these trips that he had the idea that he might be able to make something similar in England. As it happened, Roberts' computer business was located opposite the first Chapel Down winery at Burgess Hill in Sussex. According to Owen Elias, Roberts asked the winemaker David Cowderoy for vineyard advice. Cowderoy's response was 'don't bother with pinot noir and chardonnay' because they were so hard to ripen in England.

Wisely, Roberts ignored this advice. According to Lindlar, Mike and his wife Christine were already in the process of setting up a vineyard when the Mosses were getting Nyetimber going in the late eighties. At Ditchling Common in East Sussex, with some advice from Lindlar, Roberts planted 16 acres of vines with the classic Champagne varieties. The first wines were made at Lindlar's High Weald winery near Ashford in Kent, before Lindlar closed his contract winemaking business, and Roberts bought the equipment lock, stock and barrel and moved it all to Ridgeview. The venture was originally meant to be a retirement business for Mike and Christine. 'There were only four of us originally, making 20,000 bottles,' says Simon. Nevertheless, success came with the very first release – the 1996 Cuvée Merret, named after the scientist who pioneered techniques for making wine sparkle back in the 17th century.[26] It won the Gore-Browne Trophy for best English wine, an accolade that Ridgeview would go on to win many times.

The Roberts' plans for a small family project were disrupted when 'Tony Laithwaite came knocking at the door,' as Simon puts it. His business was, and still is, the country's largest mail order wine retailer, and includes The Sunday Times Wine Club. As well as selling wine, the Laithwaite family are also heavily involved with the winemaking side of the business, through their own vineyards in France and later England. Tony Laithwaite told me he'd long had the idea of making sparkling wine in England: 'Travel to Champagne and it looks a lot like the Chilterns,' he joked. 'We can do unripe grapes quite well in England.'

Ridgeview began making Laithwaites' own-label South Ridge sparkling wine – previously made at High Weald by, yes, you guessed it, Kit Lindlar: 'South Ridge was really the first remotely commercial batch of sparkling wine that I produced. It was an experiment. Someone I knew had a parcel of the right varieties [for sparkling wine] that he didn't know what to do with. I decided to carry out vinification as if we were making sparkling wine. If that didn't work I thought it would make quite a nice still wine. I showed it to a couple of people, one of whom showed it to Tony Laithwaite who said: "We like this very much; we'll have it off you."' The first vintage was 1993 but because it was aged for only twelve months as opposed to the three years at Nyetimber, it hit the market before Nyetimber's famous 1992 Blanc de Blancs. Production was very small, around 2,000 bottles according to Lindlar, but it was a great success.

These early wines are, apparently, still holding up well.

---

26 See 'Fizz Wars', chapter 8

Laithwaite told me that he recently went to see Hugh Johnson, the chairman of the Sunday Times Wine Club, and Johnson produced a bottle of the very first vintage of South Ridge. They opened it and 'it was just fantastic. How do you do that from the word go?' Laithwaite marvelled. Johnson has, over the years, been a great champion for English wines and Lindlar praised the partnership between him and Laithwaite in promoting English wine, describing them as 'the man with the vision and the man with the cheque book'.

Production of Laithwaites' South Ridge moved to Ridgeview in 1995 and transformed the business. 'They were our first customers,' says Simon Roberts. And what customers. Tony Laithwaite produced money upfront, paying a third on the pressing of the grapes, a third on bottling and a third on delivery. This is very unusual in a trade famous for its late payments. 'Tony Laithwaite was very supportive,' says Roberts. 'This is the faith he had in English wine. We owe so much to him.' It was to be the start of a beautiful friendship between the Laithwaite and Roberts families. Anne Linder described them as having 'a similar ethos – they wanted to do the right thing and find the right customers'.

Alongside South Ridge, Ridgeview was also producing Wyfold sparkling wine from Barbara Laithwaite's vineyard in the Chilterns and Theale in Reading. Anne Linder remembers planting the latter with Mike Roberts in the pouring rain. Everything had to be done by hand. Laithwaites ran an adopt-a-vine scheme where staff members would tend the vine as and when needed. Simon Roberts says it didn't quite work out like that, with Ridgeview having to send people all the way from Brighton to Reading. It was a pain, he said, but the results were worth it.

From making wine for Laithwaites, Ridgeview blossomed, all

the time maintaining its low profile. It took on Waitrose, with its Leckford Estate, and the Wine Society, the country's second largest mail-order wine business, as customers. As well as this it made wine for other English producers including the first vintages for Gusbourne, now one of the country's most prestigious brands. From a plan to produce 20,000 bottles a year, the winery now makes around half a million. In addition to its own 20-acre site in Sussex, it buys in grapes from Sussex, Essex, Suffolk, Berkshire, Herefordshire and Hampshire.

Initially, Lindlar made the wines but gradually Mike Roberts took over. His son Simon studied at Plumpton and originally wanted to stay on the viticultural side of the business but fate would intervene when he took a trip to Germany with his father and Lindlar. The trio ended up tasting wines in a monastery and one of the monks commented 'this boy has an amazing palate. You should nurture it.' And so Simon became more and more involved with winemaking, bringing a more instinctive approach to the process. His father, he says, was 'very technical but trusted my palate'. From the mid 2000s the pair began working collaboratively, with Lindlar acting as Simon's mentor. They're still in touch. At the annual Wine GB tasting in 2022, Simon mentioned to me how Lindlar had warned him of the similarities between that year and the 1976 vintage, a hot year ruined by a rainy September.

Despite the increasing interest of the press in the nineties and noughties, it was still sometimes hard to overcome English wine's image problem. Simon Roberts tells a story about being at the Sunday Times Wine Fair in London. 'We were near the Australian wines and these girls came through, tasted the wines and said, "Oh, we love Australian sparkling wine. Whereabouts are you from?"' Roberts replied: 'We're near Brighton.' 'Brighton,

Melbourne?' 'No, on the south coast of England.' The girls then tipped their wine away and walked off in disgust.

According to Simon's wife Mardi, an Australian who looks after the marketing side of the business, part of the reason for the name Ridgeview is that it sounded vaguely Australian. English wineries didn't trumpet their British origins in the way they do now – Hattingley Valley describes itself as 'unapologetically British'. Mardi explains: 'The original branding was not obviously English – that was on purpose. We wanted people to try it first.' Having said that, they did go on to adopt a branding system naming wines after different parts of London – Knightsbridge, Bloomsbury and the like – which was a bit of a giveaway.

The tipping point, says Simon, came in 2010. It was the night of the Decanter World Wine Awards and Ridgeview had been shortlisted for the World's Best Sparkling Wine trophy. The competition was two great Champagne houses – Piper-Heidsieck and Charles Heidsieck. The family were delighted just to be on the shortlist and, seeing as their table was right at the back of the mezzanine level, got stuck into the wine assuming that they didn't have a chance. Then, when awards chairman Steven Spurrier started talking about the winner, he mentioned it was a Blanc de Blancs. 'We were the only Blanc de Blancs,' Simon thought to himself. Then he exclaimed, 'Oh my God we've won!' Simon and Mike, somewhat the worse for wear, made their way to collect their trophy. 'We turned around and the whole auditorium was giving us a standing ovation.'

Suddenly, the international market woke up to Ridgeview and English wine. 'It flipped on our export switch. We had enquiries from around the world but we didn't have enough to sell.' The following year Mike Roberts would receive an MBE for services to English wine, while the award even inspired Spurrier to plant vines at his home at Bride Valley in Dorset. "Something of a poisoned chalice," Simon joked. Following the awards, Ridgeview began emphasising the origins of the wine rather than trying to hide it. By this point, the English sparkling wine scene was well established – and lively. Peter Hall had moved over to making sparkling wines at Breaky Bottom, planting Champagne varieties but sticking with seyval blanc too, with great success. He didn't have the facilities to disgorge his wines so sends them to Ridgeview for this final step. Camel Valley also moved over to being a predominantly sparkling business with some help from Jean-Manuel Jacquinot. The noughties had marked the start of the fizz boom, with several new producers coming on to the market. In 2002, hotelier Richard Balfour-Lynn planted vines at Hush Heath Estate in Kent with the aim of making England's answer to Billecart-Salmon rosé. Following on from the incredibly hot summer of 2003, there was a rash of planting across Southern England. Jenkyn Place in Hampshire was planted that year. In 2004, Andrew Weeber, a surgeon from South Africa, planted vines at Gusbourne near Rye in Kent. In 2006, the Goring family planted vines at Wiston in West Sussex and lured winemaker Dermot Sugrue away from Nyetimber. A year earlier, Ian Kellett realised the potential of the old vineyard at Hambledon by planting Champagne varieties with some help from Pol Roger. Another high-quality name, Henners, planted in Sussex in 2007, while in 2008 Simon Robinson, a senior partner at corporate lawyer Slaughter & May, put in vines at Hattingley Farm, an old Hampshire chicken farm, after hearing a segment about Nyetimber on BBC Radio 4.

In addition to Ridgeview, Mike Roberts was involved with developing the English wine industry more widely. He was chairman of EWP[27] (English Wine Producers), the trade body that promoted the sector. Nicholas Watson from Strutt & Parker estate agent, who works closely with both Nyetimber and Ridgeview, describes the transformative effect Roberts had on the industry: 'He was very much a pioneer, pulling together the industry in a more coherent way. He was one of the early, properly capitalised investors who wanted to take it seriously. To apply what they learned in professional business to making quality English wine.' One ex-Ridgeview employee, Daniel Ham, formerly of Langham and now Offbeat wines, echoes this: 'He was a great systems engineer.' While this might not sound like the most gushing tribute, Ham elaborates that it was a quality that created a solid base on which he could try new things. And Roberts' capacity for the less glamorous side of the business, such as logistics and accounting, meant that Ridgeview was the best organised producer in the country. In addition to Ham, other big names in English wine to have learned under Roberts include Charlie Holland, now chief winemaker and CEO at Gusbourne, and Josh Donaghay-Spier, head winemaker at Chapel Down.

Roberts was such a dominant figure in English wine that when he died in 2014 in his early seventies, his family did wonder if they would go on. They were 'big shoes to fill', says his daughter Tamara but, 'We ended up keeping it together as a family.' Ridgeview is now run by sister and brother Tamara and Simon – she, a former accountant, as the CEO, while he makes the wines. Doubtless there were some testing times, especially as the following year Simon suffered from endocarditis (heart infection) and spent fifty-seven days in hospital. Ham,

---

27 Which in 2017 merged with UKVA to form Wine GB

who was working at Ridgeview at the time, described how he was thrown in at the deep end. When Simon was feeling a little better, the family smuggled samples of wines in so that he could taste them from his hospital bed.

Tamara and Simon are so different in appearance – she's pale and blonde, he's dark – that many people assume they are husband and wife. When I commented on this, they laughed and revealed that Simon was adopted. His biological family were Indian Fijian. Apparently nobody had ever asked about this before. They have very different personalities too. Simon, the creative one, is warm and a little shy, whereas Tamara is all business and initially somewhat intimidating. Along with Simon's wife Mardi, who looks after marketing, Tamara's husband (also called Simon but known as Lardy, not because of his stature but because his surname is Lard) is also involved with the business.

Unlike some of the newcomers to English wine, the Roberts family don't have bottomless piles of cash. Mardi speaks enviously of the 'massive marketing budgets' enjoyed by some of their rivals. All Ridgeview's growth has to be funded by bank loans, and 'organic growth over time', according to Tamara. Their only investor was the nearby Tukker family at Tinwood who swapped lettuces for grapes in 2007. The Roberts have since bought the Tukker family out but continue to make Tinwood's wine and take some of their grapes. Tamara plans the growth of the business with detailed five-year plans. Exploring Ridgeview's accounts at Companies House shows a robust business that's profitable most years and with money in the bank – something you can't say about all English producers. Everything Ridgeview makes is ploughed back into the company. When I visited in 2022, it had just built a £2 million winery over two floors with space for over a million bottles. A new visitor centre including a restaurant called the Rose & Vine and tasting room had been stuck in development hell (it's not easy getting anything built in the South Downs national park) but did eventually open in summer 2022.

What's striking about the Ridgeview story is how much they have achieved with such little brand recognition. Partly this is to do with the nature of the business, making wines for other people, but partly it's the family's natural diffidence. It's a shame, as they have such a good story as a founding father of English sparkling wine, yet it's Nyetimber that gets all the column inches. For a small family business, there's a very corporate feel to Ridgeview. It was the only winery I visited where I signed in using an iPad and was issued with a lanyard as if I was visiting a London media conglomerate. In contrast, Chapel Down and Nyetimber, both bigger operations, were very informal. To me, the corporate image that Ridgeview projects works against them as a small family business. Tamara and Simon clearly make a great team but what the family lacks is a showman, a salesman, a Dermot Sugrue or a Mark Driver, which might be part of the reason for the oddly low profile of Ridgeview's wines.

Although they're not the ones that get the attention from the critics, the Ridgeview wines I tasted were up there with the best in the country. There is a consistency among the cheaper bottles and genuine greatness among the more expensive ones, including the famous Blanc de Blancs (a single vineyard wine made only from chardonnay grapes planted on the property in 1995). For King Charles III's first state banquet, to welcome South African president Cyril Ramaphosa in 2022, guests were first served the 2016 Ridgeview Blanc de Blancs, followed by such notable wines as Chassagne-Montrachet 1er Cru Morgeot Clos de la Chapelle, a Sauternes from Château Rieussec and Taylor's vintage port. But it was Cherie Spriggs, head winemaker at Nyetimber, who paid Ridgeview the ultimate compliment when she said, 'We see Champagne as our competition. Other English producers don't make outstanding wines year in year out and across the range. Only Ridgeview comes close.'

The close relationship between Laithwaites and Ridgeview is coming to an end as winemaking across the former's various brands has moved to Harrow & Hope in Marlow, which belongs to Henry Laithwaite, Tony's son. Perhaps now that the Roberts family are not making so much wine for other people, Ridgeview will have a chance to grab a bit more of the limelight it deserves. Ridgeview, in its quiet way, has done more than any other producer to undo prejudices over English wine by getting its bottles into supermarkets, large retailers and even exporting. Another producer, not far away in Kent, would attempt a similar thing but with still as well as sparkling wine – and unlike Ridgeview, it would do so by building an instantly recognisable label. In the next chapter we look at how a man with a background in beer marketing would turn a failing winery into a national brand.

CHAPTER 5

Not going tits up

'I was there just fending off the bailiffs'

Owen Elias

Chapel Down was in such a parlous financial state that when former CEO Frazer Thompson joined in 2001, he found bailiffs trying to repossess the photocopier. Thompson managed to dissuade them from doing so by arguing that, without it, he could not produce invoices, and therefore had no way of claiming income. By the time he retired as CEO 20 years later, Chapel Down was by far the largest producer of wine in England, with a market value of £100m.

The business is now based at Tenterden in Kent but the name comes from a vineyard on the Isle of Wight, Chapel Farm, which was planted by Anthony Pilcher in 1974. Originally, the wines were made on the island at Ken Barlow's Adgestone vineyard but when this folded Pilcher says he 'was left with the vines but nowhere to make wine or somebody to do it all'. Chapel Down was set up by a group including Pilcher, winemaker David Cowderoy from Rock Lodge in Sussex, and a charismatic money man called Nicky Branch. The name, according to Pilcher, was an amalgamation of Chapel Farm and the downs on the Isle of Wight where his vines were planted. The idea behind Chapel Down, however, was that it wouldn't own any vineyards. Instead it would operate as a wine trading business – what the French call a négociant – buying grapes, making wines both under its own label and on a contract basis, and supplying supermarkets. Nowadays, such enterprises are not uncommon in English wine but at the time this was revolutionary stuff.

Cowderoy initially made the wine at Burgess Hill in West Sussex but needed more space, something that would be a perennial problem. In 1995, the group bought what had once been Stephen Skelton's Tenterden vineyard in Kent and had subsequently fallen into disrepair; according to Owen Elias, the vines were in a terrible state, blackened and diseased. Skelton became a director of the company and all the winemaking moved to Tenterden.

Elias, having worked with Cowderoy at Rock Lodge, had been taken on as assistant winemaker. He'd come into wine after a brief but exciting career in the music business, setting up the world-music record label DiscAfrique whose biggest act was the Bhundu Boys, which ended up supporting Madonna in London in 1987. Elias found the music industry a little 'hairy', as he put it – at one gig, as Elias tells the story, someone threatened to 'chop my penis off'. It was time to get out, so he decided to retrain in what he hoped would be the more genteel world of winemaking, at Plumpton College.

In 1999, Chapel Down merged with two of the country's biggest producers, Lamberhurst Vineyards in Kent and Carr Taylor in East Sussex, forming the English Wine Group. Skelton outlines a litany of unprintable reasons as to why this was not a success. He writes, more diplomatically, in his book:[28] 'The big personalities never got on.' Elias, who by this point had become head winemaker at Chapel Down, describes it as 'a hideous episode, it really was'. Time and again when talking to Elias, the phrase 'it went tits up' recurs. A lot of the things Elias was involved with seemed to go 'tits up'. Chapel Down's early years sound particularly hard work. The company had a variety of somewhat colourful investors, and Elias recalls a lot of 'blood and financial shenanigans'. 'The whole thing was very dodgy. I wasn't privy to everything that went on. I was there just fending off the bailiffs.' Elias describes the early years as about 'dealing with shits, charlatans, liars and people with more money than you – a difficult process to combat.' He adds with triumph: 'They all disappeared in the end.'

Thompson arrived in 2001 with the company's finances in disarray. 'It was an absolute mess,' he recalls. 'We were making wine for others but people weren't collecting it. We had coach loads of visitors but nobody was buying anything.' And yet some of the wines were good. It was a bottle of sparkling wine from Chapel Down that had first pricked Thompson's interest in English wine. Then he saw an advert in a newspaper saying that the company was looking for an MD. Thompson's background was in beer. All those 'Cream of Manchester' Boddingtons' adverts were Thompson's work. From Boddingtons, he moved to Heineken and by his own account, 'became a bit of a prick'.

28 *The Wines of Great Britain*

The high life got to him. He tells a story about getting into Studio 54 nightclub in New York ahead of Heidi Klum, which sounds unlikely considering the club closed in 1980 and Klum was only born in 1973.

The Chapel Down job paid about half what he was earning at the time, and he had to put some of his own money into the business. He decided to take it anyway, if only to spend more time with the family. Coming from Heineken, 'where money was not a problem', suddenly working for a business with no cash came as a huge shock. The company was also dangerously reliant on one customer – British Airways – which made up nearly half the business. In 2003, disaster struck when the Gulf War grounded much international travel and, according to Thompson, the airline refused to honour its contract. At one point, things were so tight that Chapel Down sold Spots Farm, Stephen Skelton's old place, to Owen Elias, who stills lives there to this day. Dealing with all the various investors in the business was another strain. One of Thompson's first acts was to put Elias through a formal disciplinary process for calling one of the investors 'a very rude four-letter word', as Elias himself tells it.

Initially, they were sitting on a mountain of stock which 'wasn't shifting', says Thompson. This wasn't an unusual state of affairs in English wine. When Sam Linter took over from her father as winemaker at Bookers Vineyard in Sussex, now Bolney Wine Estate, they were sitting on ten years' worth of stock. Much of it wasn't very good and she struggled to sell it. Thompson was faced with a similar problem. Despite Cowderoy's New World credentials, most of the Chapel Down wines at this stage, even the sparkling ones, were still being made from unpronounceable German grape varieties such as huxelrebe, reichensteiner and schönburger. Making a virtue of necessity,

Thompson came up with a line of wines branded the Curious Grape to try to get people to try the wines (Elias told me that he wanted to call the wines 'Strange Fruit' but was outvoted).

Part of the problem was that although Chapel Down aimed to have very high standards in the grapes it bought, this wasn't always possible; when it came to fruit from investors, they just had to accept it, whatever the quality. Pilcher admits: 'My grapes from the island never survived the journey particularly well.' Elias recalls 200,000 litres of low-quality, low-alcohol seyval blanc that he had no home for. 'What did you do with it?' I asked. 'You have a fire,' Elias replied, his eyes lighting up, before hastily adding, 'It was not deliberate, though everyone says it was, but it solved one problem.' The company's thorough insurance policy enabled it to re-invest the payout in the winery and vineyards.

Elias has a particular hatred for seyval blanc, which he describes as 'barely a grape' for its lack of flavour and sugar. He described with relish how it pulled up and replaced with mainly Champagne varieties. Once the winery had got through its stocks of seyval, schönburger and other difficult-to-pronounce German varieties, Elias stopped buying those varieties from local growers. The bigger local growers pulled up their old varieties and replaced them with chardonnay and pinot noir but many vineyards just disappeared. As a result, plantings in England actually dropped in the late nineties and early noughties – total plantings didn't reach the 1993 high of over 1000 hectares again until 2008.

There was one legacy grape, however, that had something that the other German grapes didn't. Bacchus is much easier to say than reichensteiner, and it made an aromatic, grassy white wine that was not dissimilar to the sauvignon blancs from New

Zealand that were taking the market by storm. The variety was developed in 1933 but wasn't authorised in Germany until 1972. The first commercial planting in England was at New Hall in Essex in 1973. It wasn't an immediate hit. Veteran wine writer Oz Clarke recalls tasting bacchus in the eighties and nineties, but, 'it rarely shone; only Three Choirs regularly made something decent.' Towards the end of the century, Camel Valley in Cornwall was also having some success with it but, Clarke adds, 'It was Chapel Down that really made the name into a brand.'

The company went all out with the grape. 'We were big fans and wanted to make Chapel Down Bacchus England's answer to New Zealand sauvignon blanc,' says Thompson. Along with French varieties like chardonnay, Chapel Down began planting bacchus extensively in Kent. Initially, however, according to Clarke, 'they were mostly using Essex fruit from New Hall, who were more concerned with growing grapes than developing a brand for themselves'. Elias remembers the amazing quality of the bacchus coming out of Essex. 'We called it fat bacchus. It was from a really good site. They get the grapes ripe.'

The grape was a huge success. As well as the standard label, there followed a range of different bacchuses from single vineyards, and even a sparkling one. In the noughties, bacchus became England's USP, like malbec in Argentina. It was a godsend to producers, a grape that ripened reliably in southern England, and produced a dry wine that people liked and which, unlike a sparkling wine, can be sold a few months after the vintage, thereby avoiding sitting on capital. And more likely than not, if people bought a bacchus, it would be made by Chapel Down – Thompson estimates that Chapel Down still produces about half of England's bacchus.

The next thing to be tackled was the branding. According to Thompson, that first bottle of Chapel Down he ever tried tasted great but looked awful, so one of the first things he did was to change the branding. Chapel Down's new label appeared in 2006 in striking red, gold and black. There was nothing else quite like it. Other brands like Three Choirs in Gloucestershire and Denbies in Surrey were getting their wines onto supermarket shelves, but without Chapel Down's flair for marketing.

As well as varietal wines, Chapel Down made blends like Empire Zest (which one can't imagine getting past the naming committee these days) and Flint Dry. This last one has become something of a classic; Elias says it's still Chapel Down's best wine. It was originally made at Rock Lodge and it's clever commercially as it's partly made from second pressings of grapes which aren't considered fine enough for sparkling wine but still have lots of flavour and, crucially, less acidity. As well as buying in juice and grapes from growers, Chapel Down would buy wine for blending from other producers including Denbies, who, Elias claims, struggled to sell its own wine.

Thompson has mixed feelings about working with Elias. 'He made some amazing wine,' he says, but describes him as '"not a great manager'. Today, Elias works with his son Fergus at Balfour where Fergus described much of his job as interpreting his father's wishes to outsiders. 'It's actually quite good fun – we're both a bit odd, which works well, and I'm fluent in mumble which appears to be what dad speaks almost exclusively.' Interviewing his father was a lesson in meaningful silences and enigmatic little bursts of laughter. His inscrutability, however, was not a good fit with one Australian winemaker who worked at the business, who ended up leaving. Elias became quite animated when I brought this up, describing her as a 'backstabber' who 'would slag off my winemaking to the directors'. 'She is one of the reasons I left. I'd probably still be there but there was other stuff,' he added, before trailing off into another meaningful mumble.

Whatever his manner, Elias clearly has a good eye for talent. In 2010 he hired Josh Donaghay-Spire. A local lad from Canterbury, Donaghay-Spire graduated from Plumpton with experience working in both Champagne and Alsace, and became head winemaker in 2013. Thompson described his recruitment as a 'big risk' since Donaghay-Spire was relatively inexperienced, but it's a risk that seems to have paid off. The winemaking is now in the hands of him and Jo Arkle, 'the girl with the sparkle', as Thompson puts it, sounding a bit like Alan Partridge.

The sheer scale of Chapel Down – today it produces around 1.5 million bottles a year – means that it has problems that other wineries don't. It has to deal with grapes that don't come in at perfect ripeness, or 'shit fruit' as Thompson puts it. 'Gusbourne doesn't have these issues.' Elias, however, is adamant that after Chapel Down lost the connection with Kent vineyard Lamberhurst and East Sussex's Carr Taylor, he was strict about the sort of fruit that came in. He told me a story about one grower bringing in substandard grapes, picked days earlier and now beginning to ferment, which he just dumped straight on the compost heap. 'I had a reputation for being a grumpy git.' Nevertheless, Donaghay-Spire acknowledges that some of the wines made under the Elias regime were not as good as they could have been, mainly due to the quality of the raw materials. One wine writer and English wine expert who didn't want to be named told me that Chapel Down's sparkling wines 'don't display the abundant ripeness next to Gusbourne and Nyetimber'. Then again, they are a lot cheaper.

Things are still not straightforward for the winemaking team. Whereas Chapel Down has invested heavily in vineyards and tourism – it was full of boisterous visitors enjoying the wines and the spring weather when I visited – the winery is too small for the volume produced. Compared with the lavish facilities at Hattingley Valley or Rathfinny, it seems like a shoestring operation. Access is not great, the electrical supply is erratic and the site isn't on mains sewage, leading to problems with waste-water. Thompson says that dealing with this sort of adversity is character-building for the team. It seems, though, that Chapel Down will soon be moving its winemaking operations to a purpose-built site near Canterbury with better access and facilities, while keeping the visitor centre at Tenterden with a boutique winery on site.

Throughout Thompson's tenure, to keep the business expanding, the company needed money. Thompson says: 'You need large injections of cash if you're growing. If you don't, you're failing.' The wine business could not be less like the beer business, where you can make a product one day and sell it a few days later. Vines need two or three years before they bear fruit and sparkling wine needs a minimum of eighteen months of ageing. According to Thompson: 'It will be seven years before you see a return'. Keeping the cash coming in was Thompson's major problem. Sounding increasingly like Alan Partridge, he said in a speech delivered in 2021, 'MLK [Martin Luther King] had a dream. And now I had one. But I needed cash.' There were schemes to bring in small investors: you could lease vines for an annual fee, and in return you get wine free of duty and VAT; Chapel Down was one of the first wine businesses to get behind crowdfunding, raising nearly £3.8m in 2013, with these small investors getting a discount on the wine and becoming ambassadors for the brand. Thompson also brought in people with lots of money, like Nigel Wray, owner of Saracens rugby club and apparently Britain's answer to Warren Buffet, and global investment giant Black Rock.

The money funded a massive spend on vineyards. The business shifted from one where the winery bought in grapes, to being a major grower itself. In 2007, it planted chardonnay and bacchus near Maidstone in Kent. If you've ever driven on the A229 between the M2 and the M20, you'll have noticed that the road cuttings are white with chalk. Thompson describes it as 'the single best bit of land for growing grapes'. Donaghay-Spire agrees: 'The future is North Downs chalk,' he says. The first planting was at Kit's Coty, the site of a prehistoric megalith. This was followed by Count Lodge, Street Farm and Abbey Farms. Today, Chapel Down has 750 acres planted and 150 in the pipeline.

The results have been impressive. The first Kit's Coty sparkling wines were released in 2017. All the noise in the press was about the prestige label, Coeur de Cuvée, costing £100 a bottle and setting a new record for the price of an English wine. But the previous year's release was more important – a 2013 Kit's Coty Chardonnay that did for still wines what Nyetimber and Ridgeview had done for sparkling two decades earlier. It wasn't the first release from this vineyard – 2011 and 2012 bottlings had come out previously – but this appeared in new, elegant packaging, different from the standard red, gold and black. It was the contents, however, that really impressed. While English chardonnays had been getting better and better, this was not just GFE (good for England). This was like a grand cru Chablis, but with elegantly done oak. The 2014 was even better, while when I tried the 2017 recently next to a Meursault, I much preferred the English wine. The Kit's Coty releases put an end to Chapel Down's rather bargain-basement reputation. Thompson regrets that he didn't get into still chardonnay earlier.

Not everything Chapel Down touched has turned to gold, however. In 2018, it opened a restaurant in King's Cross in London which swiftly closed. In 2019, at the height of the craft beer boom, it opened a brewery in Ashford, with a restaurant, to produce its Curious Lager and IPA, the £1.7m start-up costs raised through crowdfunding. It never really had a chance to get going before Covid restrictions came in in March 2020, and it was put into administration before being sold to a private equity firm. Stephen Skelton was characteristically forthright: 'The beer was a fiasco – they sold the brewery for £1. Hopeless!'

When I interviewed Thompson he thought that most of what he said probably wouldn't make the book, since 'We're not an innovative producer like Tillingham or Westwell.' But for once, Thompson is being unduly modest. Chapel Down has been consistently ahead of the game with the Kit's Coty range, a sparkling bacchus and even in some years an English albariño, from Sandhurst vineyard in Kent. These are not boring wines by any means.

Thompson retired in 2021 after 20 years at the helm, during which time somehow Chapel Down hadn't gone 'tits up'. In fact, it became the biggest winery in the country. Since Thompson left, Donaghay-Spire says that it has been going through 'quite deep cultural changes'. He later emailed to say that he meant 'structural changes', perhaps because he didn't want to be seen to be criticising Thompson. The plan is to be making 2.5 million bottles by 2026. 'How big is the opportunity? Nobody knows.'

Wherever it goes, Thompson leaves behind a robust business and strong brand. He seems to take a certain delight in how many of his competitors are still to make money. In an interview with the *Financial Times*, he said: 'Margins are extremely good at the end of the cycle, and mature businesses like ours took the pain many years ago. But you've got the likes of Lord Ashcroft at Gusbourne who's seven years behind us and is burning cash furiously.' This was to be the next chapter of English wine, as massive wealth entered the industry; people for whom dropping a few million on a winery is pocket change. It was the rise of the money men.

CHAPTER 6

Money men

'I might be quite demanding. I demand a lot of myself. I am a perfectionist'

Eric Heerema

In 2018, the team from The Pig hotel group put on a dinner with winemakers to celebrate the imminent opening of its newest venue, in a converted Jacobean manor house near Canterbury called Bridge Place. Bridge Place has had a chequered history. In the sixties and seventies it was a music venue hosting bands like Led Zeppelin and the Kinks. Then it fell on hard times and became an over-30s nightclub, notorious for what our cab driver Kevin called 'grab-a-granny' nights. The Pigs are a small but upscale group of country hotels that major on food and wine. The one in Cornwall sells over 10,000 bottles of Camel Valley sparkling wine a year so you can see why the Kentish winemakers would want to attend.

did say that you need to wait at least seven years before you make a return on your investment, and that building a premium wine brand is the work of 20–30 years. 'I wish I'd started when I was ten,' he jokes.

2002, when Balfour was set up, marked a turning point for English wine, according to Nicholas Watson at Strutt & Parker estate agent, who counts Ridgeview and Nyetimber among his clients. 'English wine changed around twenty years ago. Previously, British viticulture was populated mainly by enthusiast hobbyists, and professional businesses were in the minority. Nowadays they are much more grown-up, properly capitalised and properly business-like. The wine being produced now is remarkable.' The noughties saw the arrival of massive amounts of money into English wine from a variety of other successful businesses – watercress barons at Exton Park, lettuce magnates at Tinwood, and hoteliers in the form of Balfour-Lynn. But most notable of all were the people who took other people's money and did extremely lucrative things with it that I don't quite understand.

The financial boom in the 1980s had brought new money into English wine but the influx of investment at the turn of the century was on a different scale. It was the kind of money that small producers like Bolney, which had long struggled to attract investors to fund its expansion, can only have looked on with envy. The steady rise of the stock market that began in the late nineties, was briefly interrupted by the 2007 financial crash and is now finally coming to an end as I write this, made certain people – particularly those involved in finance – stupendously rich. Combine that with unusually low interest rates – something that is again coming to an end – and there was a lot of cash swilling about in the first part of the new century, looking for a return. Many of these City types were wine lovers so where better to put their money than in a vineyard?

Take Mark and Sarah Driver. He was a hedge fund manager with a stint at an investment company in Hong Kong, and she was a lawyer. Between them, in 2010, they bought the 600-acre Rathfinny farm on the South Downs, just three miles from the sea. The farm gets its Gaelic name from an Irishman called Finny who used to own the estate (it was originally called the less romantic-sounding Pinchmedown Farm). The vines were planted in 2012 and the first harvest was in 2014. The Drivers must have been pretty successful in their chosen fields as the scale of their investment in Rathfinny is gigantic, with the vines stretching for as far as the eye can see over perfect chalkland. It makes a change from a muddy field in Kent.

When the world woke up to the fact that good sparkling wine could be made in England, all the talk was about chalk soil, even though the original plantings at Nyetimber were on greensand. For those with deep pockets the dream was acre after acre of purest chalk. What is it about chalk that gets wine types so excited? Yes, it looks cool, all that bright shimmering white, but most importantly it makes people think of Champagne. Some of the finest vineyards of the Côte des Blancs and the Montagne de Reims are on chalk (though there are also great wines there made from clay and gravel). Ultimately, the chalk-being-the-same-soil-as-Champagne story has done much to sell English wine.

The other beauty of chalk is that it sucks up water like a sponge, stopping the vines' roots from getting water-logged. According to Charlie Holland, CEO and head winemaker at

Gusbourne, which has vineyards on both clay and chalk, grapes grown on chalk tend to start ripening sooner but because chalk doesn't retain the heat so well, they finish ripening later and have higher acidity. In contrast, clay takes longer to warm up but once it's warm, the grapes ripen faster, producing richer, fruitier wines. The Gusbourne style is a mixture of the richer clay and the fresher chalk wines. Chapel Down has bet the farm on chalk, with 100 acres of vines going in across the North Down at vineyards such as Kit's Coty. The Drivers too went all out for chalk. According to Mark, 'We sent the soil away for analysis in France and the lab thought it was from Champagne itself.' This is a not an uncommon story in English wine. Stephen Duckett from Hundred Hills in Oxfordshire had exactly the same experience.

There's some debate, however, as to whether chalk actually makes that much of a difference. For many producers, other factors such as microclimate, drainage and elevation are much more important, especially in a climate as marginal as England's. As we saw previously, you can plant vines on a load of rubble, as long as it's in the right place, and make great wine. Nicholas Watson from Strutt & Parker says that the most important thing is for the soil to be free draining. 'If soil is free draining, that's half the battle. The soil doesn't want to be too rich and fertile. If you put a vine in fertile soil it will go crazy and produce lots of foliage rather than lots of grapes.'

Many thought that the Drivers were making a rather expensive mistake when they invested so heavily in Rathfinny which, being so close to the sea, is very windy. One of the most vocal naysayers was arguably the foremost expert on vineyard sites in England, Stephen Skelton. Mark Driver, however, thinks that

Skelton was only carping because he wasn't consulted on the project. According to Driver, Skelton called him and asked 'Why haven't you consulted with me?' 'I had never heard of him,' is Driver's response.

Skelton isn't the only one who has doubts. Oz Clarke writes:[30] 'The chalk slopes near Eastbourne face south, but they do get a belting from the winds coming off the sea only a couple of miles away. The vines planted so far need 10,000 trees to be planted as windbreaks.' Driver admits that the wind is a problem. 'Winds above 10mph shut down plants. Not good.' The high wind has meant that the trees planted as windbreaks haven't grown as quickly as he hoped. Here, even the windbreaks need windbreaks, it seems. The wind, however, does have an upside as it dries out the vineyard, meaning that they don't have the problems with mildew that other sites have.

From the stories about him, I expected Driver to be somewhat full of himself but, disappointingly, he turned out to be disarmingly charming and open when I met him. He described starting a wine business as the 'double-double' rule: everything takes twice as long as you think it's going to and costs twice as much. It's not only the size of the Rathfinny vineyard that impresses. With its three gigantic £60,000 Coquard presses, the winery is on a huge scale. The aim is to make 500,000 bottles per year. No expense is spared and the wines are not released until the Drivers are happy with them. 'You need age to produce quality,' Driver argues. The flagship wine, inspired by Bollinger, is the Blanc de Noirs; as I write in 2022, the current vintage available is 2015. Think of all the money tied up in bottles not yet released. Driver thought he would break even in 2023,

---

30 *English Wine*, Oz Clarke

though this doesn't take into account the estimated £10m he has put into the project.

Men like Driver may be wine lovers but they are also hard-headed businessmen. As another wealthy vineyard owner, Stephen Duckett at Hundred Hills, told me, 'It's about yields on capital.' Duckett made a fortune in tech in the 1990s and 2000s but, tired of being the oldest person in the room, he returned from California and sunk money into a vineyard. He wouldn't tell me how much he had spent but it can't be far behind what Driver has put in despite Hundred Hills being a much smaller concern. For Duckett, the point is to build a 'profitable family business'.

The Drivers began building a brand even before they had wine to sell. They put money into Plumpton College via the Rathfinny Wing: good for English wine and great publicity. Their biggest coup, however, was securing a PDO (Protected Designation of Origin) for Sussex which had been in the pipeline from 2015 and was finally approved by Defra (Department for Environment Food and Rural Affairs) in 2022. This means that bottles that pass certain criteria get to carry Sussex branding, rather like products such as Stilton cheese or Jersey royals. There is already a PDO for English wine and for English sparkling wine. And Bob Lindo has attained a vineyard-specific one for Darnibole Bacchus made at Camel Valley in Cornwall.

The Sussex PDO is an attempt to answer to the perennial problem of what you call English fizz.[31] Mark Driver says: 'We believe that the name Sussex will become synonymous with high-quality sparkling wine, so that when you go into a bar in London, New York, Beijing or Tokyo you will be asked 'Would

---

31 See 'Fizz Wars', chapter 8

you like a glass of Champagne or a delicious glass of Sussex?'
According to Driver, the idea was originally Mike Roberts',
though most wines produced by Ridgeview use fruit from
outside Sussex so are not eligible. Nyetimber too is not eligible,
so the PDO is missing the historic county's two biggest producers.
Sussex sparkling wines can only use Champagne varieties,
including some of the lesser-known ones, while still wines can
use a much wider selection of varieties including some hybrids.[32]

Not everyone shares Driver's vision though, and the Sussex
PDO has come in for a fair degree of criticism. Henry Laithwaite
from Harrow & Hope near Marlow has argued that the industry
doesn't yet understand enough about its sites to draw a line
around regions. 'It's not there yet,' he says. 'Give it 50 years.'
Tamara Roberts, daughter of Mike Roberts, won't be applying to
join the PDO even though some of her Ridgeview wines are
eligible. She doubts the term means anything to the customer
and feels it would be better for such a young industry to be
united in its approach.

But perhaps the most eloquent and vocal critic has been
wine writer Victoria Moore in the *Daily Telegraph*: 'Are there
any entire counties whose grapes could be recognised in a
blind tasting? No . . . because county boundaries are political
and agricultural boundaries are determined by soil type, slope
elevation, orientation, proximity to water, and microclimate.'
Moore thinks that such premature PDOs 'imperil the development of the wine industry in two ways. First, creating regulations
too early risks stunting the experimentation that might create
even better wines. Second, it lets down the drinker and the
whole English wine system, both of whom deserve a more

---

32 See 'Grape Expectations', chapter 12

coherent and collegiate approach when it comes to defining large PDOs.'

She has a point. It seems odd to tie Sussex sparkling wine so closely to Champagne if you're trying to produce something unique. The PDO does also cover still wines but the rules are much less strict than for sparkling. In fact, the number of grape varieties allowed in the still category is so broad as to make the concept of a distinctive Sussex wine somewhat meaningless. All in all, it's something of a blunt instrument, but it's certainly got people talking, which is part of the point. And as Laithwaite said of Driver: 'He's poured millions into an asset so he has to protect it. It's pure marketing.'

Driver seems genuinely hurt and surprised about the vehement opposition to his scheme. He sees defining regions as the logical next step for English wine, something that has also happened in the New World. Australian wines, for example, used to be sold on brand and grape variety before the country set up its own geographical indications like Margaret River and Coonawarra. And as Driver points out, Burgundy and the Languedoc are both historical, political constructions rather than geographical, and 'not [made up of] the same soil or climate but still regions'. Though say the word Burgundy and you think of a distinctive style of wine, something which is true, to a lesser extent, of the Languedoc. Driver sees the PDO more as a guarantee of quality than an indicator of style, with stringent testing in order to qualify. Grapes will have to be a certain ripeness and the finished wines will have to pass a judging panel and be scored at least 16/20. So far, not many producers have signed up to the scheme, with only Bolney and Rathfinny among the large producers involved. But other counties are watching with interest. An article in the *Financial*

Times quoted Taittinger's UK representative, Patrick McGrath of Hatch Mansfield, as suggesting that the eight producers that make up the Wine Garden of England are considering launching an application for their own Kent PDO.

Even Driver's harshest critics wouldn't doubt the quality of the sparkling wines he's producing. But he has been less successful with still wines. He planted five acres of riesling in 2012 and to say it didn't flourish would be an understatement. The best year was probably 2017, when he managed to get the sugar levels up to around 11% potential ABV – not bad at all – but the acidity levels were off the scale and it tasted greenly underripe. 'We didn't release the wine,' says Driver. 'I wasn't happy with it.' Five years after planting, he pulled the riesling out and replaced it with chardonnay and pinot gris. It wasn't completely wasted though. A nearby gin producer, Silent Spring, distilled it into a brandy with various other winery leftovers. Aged for three years in sherry casks it tasted quite superb when I tried it after lunch at Rathfinny. Maybe there should be space in the Sussex PDO for a brandy.

People with deep pockets like Driver can make mistakes like this and it doesn't affect their main business. Wine writer Tom Hewson described failed wines as 'cannon fodder', taking the hits for the industry as a whole. At the larger wineries, wine-makers often have the freedom to make experimental batches. At Balfour, the main business is sparkling wine. There's the famous rosé as well as more everyday offerings like Tesco's Finest English Sparkling Wine. But winemaker Fergus Elias also produces a dizzying array of limited-edition wines where he gets to experiment with different plots, grape varieties and yeast strains. These culminated in 2020 with the creation of red wine dubbed The Gatehouse made from the finest plots of pinot noir.

With its deep colour and intensity of fruit, it tasted like something from Austria rather than Kent. It had a price tag to match, £60, but even then it probably doesn't make much money for Balfour. It's the wine equivalent of a concept car, pointing to what might be possible in the near future. When I ran into Elias at the annual Wine GB tasting he was enthused about the albariño he had made the year before, though what he really wanted to make was an aligoté, a Burgundian white grape, which nobody grows in England . . . yet.

If you're rich, you can afford to be a perfectionist. At Hundred Hills in Oxfordshire they sell off around a third of the wine produced which is not up to scratch. They also sometimes sell off bottled wines that haven't met the Ducketts' exacting criteria, which will then be slapped with the label of another brand. According to Rupert Crick, hospitality manager at the vineyard, this is not unusual in the English wine world, though he wouldn't name any other names. Hundred Hills is also pushing the boundaries when it comes to technology, working with a team of scientists on an AI system where the health of vines can be monitored from a satellite just by assessing the colour of the leaves. Slightly more down to earth, at Roebuck in Sussex, they use an app called Sectormentor which monitors the health of vines, soil and fruit on a block-by-block basis, building up a vintage database so producers can better understand the terroir. The idea is to acquire, over the course of a few harvests, a knowledge bank that has taken traditional regions generations to build. Money allows owners to pay the best people too. Rather than attend a winemaking school like Plumpton College or Roseworthy in Australia, Stephen Duckett was taught personally by the late Michel Salgues of Champagne Louis Roederer.

The money men haven't just established new estates, they've also gobbled up existing enterprises. Gusbourne, established in 2004 by a South African doctor called Andrew Weeber who was famous for fixing Paul Gascoigne's knee, sold up in 2013 to a consortium led by businessman and politician Lord Ashcroft for £7 million. But the biggest prize of all was Nyetimber, bought from owner Andy Hill by Eric Heerema, a Dutch shipping magnate and entrepreneur in 2006 for £7.5m. Heerema had already dipped his toes into wine with a small planting of vines at his property on the South Downs, but this was a whole different proposition. On taking over, Heerema described the winery as 'lots of enthusiastic people but a cottage industry undertaking with no real sales team and no marketing. There were some beautiful vineyards and some beautiful land where I figured I could plant. But it all required a lot of work.' It also required a lot of money. Stephen Skelton[33] estimates that Heerema has put in at least £50m and probably more. Nyetimber's 2021 accounts showed that it owed nearly £91m, up from £82m the previous year. This isn't the problem that it sounds though, since most of this money is owed to Nyetimber Wines, a Jersey-based company wholly owned by Heerema. He thinks that profitability should be possible 'in the next three or four years. In the long run we should become very profitable.'

With this kind of cash, Heerema set about fulfilling the previous owners' ambitions and turning Nyetimber into a consistent, globally recognised sparkling wine brand. His single-minded approach has not always made him popular. Former head of sales James Mason won a court case against Nyetimber

---

33 *The Wines of Great Britain*

for unfair dismissal in 2017. He said at the time:[34] 'I was wrongly accused of being lazy, incompetent and lying. The judge ruled that I was wrongly dismissed on alleged grounds of gross misconduct and negligence. If I had not stood up to Mr Heerema, he could have continued to ruin drink industry careers. During my eighteen-month employment period at Nyetimber, thirty people either left or were fired by Mr Heerema.' During the tribunal a story emerged that The Savoy Hotel had refused to stock Nyetimber following a now notorious incident. According to *The Times*: 'Eric Heerema, proprietor of Nyetimber Wines, is alleged to have swept a table full of Savoy-branded glasses onto the floor and shouted at a bartender during an event at the Burlington Arcade in London. The incident is said to have prompted the head of food and drinks at The Savoy to write to Nyetimber saying he would no longer do business with the vineyard because of Mr Heerema's "appalling behaviour".' Meanwhile, former head of marketing Louise McGuane commented on Linkedin: 'Whilst there I saw many great young people in the business either dismissed or set against by the owner for no other reason than personal feelings rather than performance. After a few months, I came to view my role as one of protecting my team of young professionals from the owner's personal opinions of them.'

With some trepidation during our down-the-line interview, I asked Heerema whether such coverage was fair. He replied: 'I might be quite demanding. I demand a lot of myself. I am a perfectionist. I am very hard-working. Our ambition is very hard to achieve and we need to demand a lot from ourselves. There have been people who came in, even some people who have

---

34 *Harpers Wine and Spirit*

some experience in the drinks industry, who might not have fitted in with what we were doing. I guess I am seen as difficult and also seen as straight-talking. I am a Dutchman. We Dutch are a little over-direct.' He then pointed to the strength and longevity of his relationship with his winemaking team, husband-and-wife Cherie Spriggs and Brad Greatrix: 'We have an incredible bond of trust. I have that with other people in the company. I am very proud of the team we have.' The couple are originally from Canada and emailed Nyetimber on spec to see if the company was looking for a new winemaker. It turns out that it was, Dermot Sugrue having just left for West Sussex vineyard Wiston Estate. 'They had no idea I was looking internationally for a highly experienced winemaker,' says Heerema. 'I intended only to take one but they made it very clear that they came as a couple.'

It has proved a great working relationship. The couple have been with Nyetimber for 16 years now. I tasted my way through the range with Spriggs just after she had come back from maternity leave. There wasn't an ordinary wine among them – from the entry-level non-vintage Classic Cuvée right up to the ultra-expensive 1086 wines, named after the date when Nyetimber was mentioned in the Domesday Book. It's no wonder that Spriggs became the first person from outside Champagne to be crowned Winemaker of the Year at the 2018 International Sparkling Wine Challenge.

Production is now huge by English standards – up from around 20,000 in the 1990s to around a million bottles in the excellent 2020 vintage, with plans to double this by 2030 – though this is still less than a small Champagne house like Pol Roger. Nyetimber's home might be a medieval manor house but its winemaking now takes place on an industrial estate near

Crawley – not as glamorous, but much better for logistics. The set-up is quite extraordinary with a high degree of automation – it's truly a factory of sparkling wine. To feed the machines, Nyetimber has plantings not just in Sussex, but in Hampshire and Kent as well, with 420 acres in total and a continual search for more suitable land. It is also sitting on over £30m worth of stock.

The appeal of owning an English vineyard for people with deep pockets has driven up the price of grapes and land. According to Nicholas Watson from Strutt & Parker: 'The bottom line is that only a very small percentage of land is suitable for wine production, a very, very small proportion of which changes hands on the open market each year.' As a result, it's a question of persuading people to sell, which means potential owners needing even more money, he says. 'Most of the work we do is helping clients identify the suitable land, researching who owns it and negotiating privately with them. You often have to offer a premium to entice them to sell.' Vineyard land has risen largely because landowners know that people looking for it tend to have money. Good vineyard land in the south of England is going to cost about £15–25,000 an acre.

This might sound like a lot of money but in Champagne, similar land might cost £500,000 per acre. You can see why Champagne houses – and those from further afield – might take an interest in south-east England. And many of them are doing just that . . .

CHAPTER 7

Foreign affairs

'We weren't interested in buying an existing estate. Or doing what Pommery had done and collaborating with an English winery'

Patrick McGrath

In 2022, Clément Pierlot, cellar master at Champagne Pommery, caused a wave of indignation in the English wine industry when he said in an interview:[35] 'It's not justifiable for an English sparkling wine to cost as much as, or more than, a bottle of Champagne, with its 300 years' history and its reserve wines.' Twenty years ago, this would have been a statement so uncontroversial that there would have been no need to make it but what makes it such a bizarre thing to say these days is that Champagne Pommery has, since 2018, produced an English

---

35 *Drinks Business*

wine, called Louis Pommery. Is Pierlot really saying that no English wine produced by Pommery will ever cost as much as a bottle of Pommery NV Champagne, currently on offer at Majestic for £34.99? Is England, with its low yields and high labour costs, really the place to be making a budget Pommery? M. Pierlot, however, will be pleased to note that a bottle of Louis Pommery is currently selling for a mere £27.99 at Majestic.

Pommery was the first Champagne house to release an English wine under its label, though how much it was a 'Pommery' wine is open to debate. It was made by Hattingley Valley to Pommery's specifications, using grapes bought in from Kent, Sussex and Hampshire (Pommery only planted its own English vines in 2017). Long before this, since the early 2000s, there had been rumours in the press of Champagne houses scouting for land in England. The French coming to England to make wine was too good a story to miss, even if there wasn't much substance to it. And not just Champagne houses; over the years Michel Chapoutier from the Rhône Valley, Bernard Magrez from Bordeaux and a lone American, Randall Grahm from California, have all been rumoured to be interested in making wine in England. It's the big Champagne houses, however, that seem to have been the most serious. Duval-Leroy has variously been linked with Wiston in Sussex, Squerryes in Kent and Steven Spurrier's Bride Valley in Dorset but nothing ever came of these relationships. Billecart-Salmon got the furthest down the line in a joint project with wine merchant Berry Bros & Rudd. John Atkinson, now with Danbury Ridge in Essex, was scouting for suitable chalkland and said that everything was ready to go when the financial crash of 2008 led Billecart-Salmon to pull out. Which in retrospect looks like a terrible decision.

Four years previously, one lone Champagne-maker had already taken the plunge and bought a little bit of England; Didier Pierson, now at Champagne Frerejean Frères. His involvement in English wine dates back to 2004 when he planted four hectares of grapes on the South Downs in Hampshire. In a nice twist to the usual story, he was inspired by Mike Roberts of Ridgeview. Pierson was married to an English woman, Imogen Whitaker, at the time so wanted to do something over here. In an interview[36] Pierson said: 'I knocked on the door of a farmer and said, "I want to plant a vineyard on your land," and he looked at me as though I was mad.' Nonetheless, he ended up going into partnership with a farmer and the result was Meon Hill. The first harvest was in 2007 and the first wine was released in 2009. Following the break-up of his marriage, Pierson moved back to France and, in 2014, Meon Hill was taken over by Hambledon.

The first of the big boys to buy land in England was Taittinger. Such was the level of press interest in Champagne houses investing in England that when Patrick McGrath from Hatch Mansfield and Pierre-Emmanuel Taittinger were scouting for land here, they made sure they spoke English, so no one would get wind of the Anglo-Gallic alliance. (Maybe the two of them should have dressed up like the Thompson Twins from Tintin in a bid to look more British; Taittinger is so outrageously French in his mannerisms that it's hard to imagine him passing for anything else.)

The official announcement of a partnership between Taittinger and Hatch Mansfield came in 2015. They had bought 120 acres of prime fruit-growing land near Chilham in Kent.

36 *Drinks Business*

The family they bought from were the Gaskains, whose ancestors came over the last time the French were interested in acquiring large amounts of English countryside, during the Norman Conquest. The venture is called Domaine Evremond, named after Charles de Saint-Évremond, a French diplomat and rogue who in exile from France is credited with introducing the wines of Champagne to the court of Charles II of England. His parties, known as *petites soupers*, were notorious for scandalous behaviour, and Pierre-Emmanuel Taittinger seems to regard him as a kindred spirit. Despite this light-hearted style, Taittinger takes the ties of friendship between England and France very seriously. People have long memories in Champagne. There are war graves throughout the region of British and Commonwealth soldiers who died fighting for France. But it's also about the close friendship between the two men involved. As McGrath said, 'Since 1988, we have been joined at the hip.'

In 2017 the vines went into the Kent ground on a blustery cold May morning with the country's press in attendance[37]. Of course, it would be a few years before the vines would produce grapes and many more years before that wine would be released. We would have to be patient before we would get to taste Domaine Evremond. Unlike Pommery, Taittinger didn't want to release a wine that wasn't made from its land and its grapes. So when I revisited in 2022 there was nothing to drink except pints of Shepherd Neame at The Woolpack pub. The domaine had certainly changed a lot over the following five years. Rather than a muddy field, there were vines and as it was July, there were grapes galore. It looked like it was going to be a superb harvest, though according to McGrath they were desperate for

---

37 See introduction

rain, worried that the lack of water might have an effect on the crop. McGrath is one of those Englishmen that you can immediately recognise as ex-army. Tanned and lean and in his early 60s, one can imagine him wearing a keffiyeh and fighting in the deserts of North Africa.

He said the idea for Domaine Evremond came from a 'friendly discussion with Pierre-Emmanuel. We tasted some wine and decided to do something together. The aim was to create something like Domaine Carneros,' the Taittinger family outpost in California. It was important that they do this 'from scratch and long term. We weren't interested in buying an existing estate, or doing what Pommery had done and collaborating with an English winery.' This is very much a long-term investment. They bought the land in 2014 and the first bottle won't go on sale until 2024. The first planting in 2017 was 24 hectares, with more vines going in the ground in 2019, 2020 and 2021, reaching 52 hectares in total. Around 130,000 bottles were produced in the first two vintages, 2019 and 2020, but the plan is to make around 200,000 a year. The 2019 vintage was made at the nearby Simpsons winery before Taittinger built a temporary winery on site. It's in an old farm building and quite rudimentary when you compare it to the grandiose Taittinger HQ in France.

The reason they're making wine in such cramped conditions is because of the usual British problem in getting planning permission. Apparently one local lady has held things up for years with objections that keep getting thrown out but have delayed the process of building a proper winery considerably. McGrath describes her as a 'serial objector'. He adds, 'she cost us a fortune as construction costs went through the roof.' Entrenched nimbyism is another reason why it pays to have deep pockets in English wine; without money, nothing would get built. Now though, work is under way and it's clear from my visit that Domaine Evremond is not doing things by halves. Hard hats and high viz on – safety first – McGrath led me towards what looked like a chalk quarry. It was a huge hole in the ground, nearly 30m deep, with a treacherously steep path winding down to the bottom. It was a very hot, sunny day, and this hole was pure white chalk. At the bottom of the hole I had to squint from the glare (everyone else had sensibly brought sunglasses). It felt like the set for a low-budget BBC sci-fi series from the 1980s. I could almost imagine Sylvester McCoy or Tom Baker as Doctor Who doing battle with the Daleks. The planned winery will be two-thirds underground so rather like the chalk cellars of Champagne it will require no cooling. According to McGrath, it took eight weeks to dig out and the construction work should be finished in 2023, just in time for the harvest in October.

Though the scale of the winery is impressive, they're not planning on turning Evremond into a tourist attraction. There won't be a restaurant or bar on site. Visitors will be able to come and taste and buy the wines but will then be encouraged to visit local restaurants and pubs. Possibly as a result, they're clearly popular in the local community – everyone at The Woolpack seemed to know McGrath. It would have been very easy and understandable as a Champagne house to stand somewhat aloof of English wine, but Domaine Evremond has gone out its way to fit in with other producers. It's a founder member of the Wine Garden of England group, and at the press event in 2017 served local wines alongside Taittinger Champagne and its Californian sparkling wine.

We went for a stroll around the site including a visit to the field where I planted my vine back in 2017. Apparently on the press day there had been a mix-up and some journalists had planted chardonnay where they should have planted pinot noir. All I can say is sorry, and I hope it doesn't affect the quality of the wine. Where before there was a muddy field, now there are rows and rows of vines. But it's not a monoculture – there are still patches of land containing fruit trees. There are also wild flowers and a pond, and the team is working with a local ecologist to make sure the site is as biodiverse as possible. When I asked McGrath whether this actually helps the vines he said 'no' and then added that he wasn't sure, though Rupert Ponsonby from the PR agency, who is also an organic farmer in the Cotswolds, was sure that it did.

Though the Gaskain family sold the land to Taittinger, they have stayed on to look after the grapes. According to Mark Gaskain, 'As the fifth largest fruit grower in Britain, we bring local knowledge, and challenge them [Taittinger] just as they challenge us.' In 2019, the Tattinger viticultural team lost some of the crop to mildew as it had been spraying preventative treatments 'based on a continental climate rather than the damper maritime climate which has more disease pressure,' Gaskain says. The two teams are now working well together. Taittinger's own vineyards in Champagne are organic and Evremond's eventual aim is to follow suit, but for now the vines are farmed conventionally using synthetic fungicides – important in England's damper climate. Most of the weeding is done mechanically rather than 'mullering with herbicide' as Gaskain puts it. 'We don't mind weeds as long as they're not harbouring anything. We don't want a wasteland.' Gaskain's father described 2021, with its cool, damp summer, as the most challenging year he had seen in over 50 years of growing fruit in Kent. Yet

Domaine Evremond had 'a cracking year', only a bit down on 2020, something McGrath puts down to a good site and good husbandry.

There are other things that are done differently in England. The vines are planted further apart so that they get as much light as possible. Gaskain refers to plants as 'farmers of light – the leaves are solar panels'. Less crowding also reduces the chance of a fungal infection. Removing excess leaves to allow the grapes to air is another important part of keeping them disease-free. Due mainly to the cooler, damper weather, yields are approximately a third less than in Champagne. Gaskain thinks you should be able to get 3.2 tonnes per acre, compared with 5 tonnes per acre in France.

Yields might be lower but the land is significantly cheaper than in Champagne where a hectare (2.5 acres) of prime vineyard could set you back £1 million. Compare that with £20–30,000 for an equivalent plot in the South Downs. But it's not just about cheap land. There is also the worry that with climate change it might not be possible in the coming decades to produce first-rate sparkling wines in Champagne. To make wines via the méthode Champagnoise,[38] you need grapes with high acidity. This is part of the reason why up till now nobody has been able to challenge the Champenois for sparkling wine. The climate in Champagne is unique in allowing the grapes to ripen just enough but not too much. Combine that with an industry that's been doing it for hundreds of years with unrivalled marketing power and money, and you have a recipe for why Champagne has become so successful.

Over the last 30 years, temperatures in Champagne have

---

38 See 'Fizz Wars', chapter 8, for a fuller explanation

risen by approximately 1°C. 2019 saw the mercury go above 40°
and harvests are getting earlier. Ever since the incredibly hot
2003 vintage, they now often take place in August (2013 was a
recent outlier, with the harvest in October, just how things used
to be). At the moment, the warmer weather is working out quite
well for Champagne, though in some years soaring temperatures
have led to sunburnt grapes. The wines are better than ever, and
the riper grapes mean that Champagnes don't need the addition
of so much sugar. But the worry is that if current trends continue,
the grapes will get too ripe and lose the all-important acidity.
Thibaut Le Mailloux from regional body the CIVC (Comité
Interprofessionnel du Vin de Champagne) says: 'It's happening
very fast. The challenge is to be able to still produce brilliant
wine in ten, twenty, thirty or fifty years' time.' Currently the
climate in south-east England is similar to that of Champagne
in the 1970s, so you can see why Champagne houses are keen
to invest here.

Despite the lower prices of English chalkland, it's still not
cheap buying the right vineyard land. McGrath says: 'We want
more chardonnay and are looking for land but people hear the
name Taittinger and the price goes through the roof.' The
plan is to create a non-vintage (a blend of multiple vintages), a
rosé and a premium vintage wine. Most of the sales will take
place in England where there seems to be plenty of demand, at
least at the moment, but McGrath reckons that they could
export between 10–15 per cent, taking advantage of Tattinger's
'ready-made distribution channels'. The English wine world is
waiting with bated breath to see what the wines will be like
when they are released in 2024. The way Taittinger has handled
the PR for Domaine Evremond has been masterful in terms of
building up anticipation, but can the wines live up to all the

hype? I have a feeling that anything less than ascent straight into the front rank of English sparkling wines alongside Nyetimber will be seen as a failure. McGrath commented: 'We know that the expectation surrounding the release of the first bottle will be very high but we also firmly believe that it will be very good. Will it be as good as it will be in five years? Clearly not as there will no reserve wine in the blend and the vines will be very young. But we're confident that the UK trade will take that into consideration.' He added: 'And if you never try, you never succeed!'

Taittinger's great rival in English wine, Pommery, also planted its first vineyard in 2017 – thirty acres at Pingelstone in Hampshire, with the first harvest in 2020. Pommery will release the first Pinglestone Estate wine in 2024 at the same time as Domaine Evremond, making for an interesting comparison. Because of the inevitable planning problems, Pommery's own winery won't be up and running until 2024 so the first Pingelstone releases will be made at Hattingley Valley. Pommery has more land which it plans to plant, and eventually Louis Pommery non-vintage will use only Pommery grapes from Hampshire.

McGrath thinks it's only a matter of time before a global wine and luxury goods giant like Moët Hennessy, which owns Moët et Chandon, Veuve Clicquot and Hennessy Cognac, decides that it's time to add an English winery to its portfolio. In December 2022, however, in an interview with the *Daily Telegraph*, chief executive of Moët Hennessy Philippe Schaus denied that his firm had any interest in acquiring land in England. 'We're not doing that,' he said. 'We will keep investing in Champagne.' He went on to say that it wasn't helpful to compare the two regions: 'Champagne is not just about average temperatures. The soil is very particular. The craftsmanship is very, very particular. There's an enormous effort to keep this craftsmanship and develop it. There's much more to a category like Champagne than just latitude.' It does sound like someone at Moët Hennessy has really thought about it.

When an international wine giant did buy up an English wine estate, it wasn't from the expected place, France, but Germany. In January 2022, it was announced that Freixenet-Copestick had bought Bolney estate in Sussex for an undisclosed sum. The company is the British arm of Henkell-Freixenet which, though headquartered in Germany, is best known for its ubiquitous Cava in a black bottle, Freixenet Cordon Negro. It also makes huge volumes of Sekt, German sparkling wine, has wine interests in Australia and owns one of Germany's most prestigious wineries, Schloss Johannisberg.

The deal rather had the English wine world scratching its collective head. McGrath said that 'The industry thought it was an odd choice. Bolney is not a big brand.' And nor, despite some occasionally excellent wines, is it a prestigious one. There are rumours that the German multinational had got quite far down the line in acquiring a much better-known producer before the deal fell through and it snapped up Bolney instead. But Bolney does have a rich history. Originally known as Bookers Vineyard, it was founded at Bolney in West Sussex by Janet and Rodney Pratt in 1972. They were part of the pioneering generation which included the Barnes family at Biddenden and Peter Hall at Breaky Bottom. Their daughter Sam Linter joined the firm in 1995 and, soon after, the name was changed to Bolney.

Today, Bolney makes everything from cheap – for England – fizz, to some high quality saignée-method sparkling rosé, and red pinot noir that was once the best in the country but has been somewhat overtaken by developments at Gusbourne in Kent and Danbury Ridge in Essex.[39] Even before the takeover, Bolney was changing direction. It had endured a difficult 2020, when a large investor's business collapsed during Covid, forcing the winery to sell its wonderfully named Pookchurch vineyard to keep the business afloat. But Linter found a new investor straight away and a rebrand was in the pipeline when Linter relinquished winemaking duties to a young South African, Cara Lee Dely. After something of a baptism of fire (or rain) in 2021, Dely was set for better things in the more promising 2022 vintage.

The takeover happened in the same year – the 50th anniversary of the company. The family no longer own any shares but Linter is staying on as managing director.[40] It had always been a struggle funding growth at Bolney, she explains: 'As a family, we have taken this as far as we can with our resources. We reached the end of what we could do on our own.' Linter, who is the current chair of industry group Wine GB, thinks that Bolney was somewhat a victim of its own success: increasing production of sparkling wine means a huge investment in stock. 'We couldn't keep up with demand so we couldn't do long lees ageing.'[41] She talked to a few people before selling and felt that Henkell Freixenet represented 'the right people to preserve the DNA of the company'. There were worries in the industry that the plan was to make England's answer to Freixenet's highly

39 See 'Eastern Promise', chapter 13
40 Just before going to press Linter announced that she had left Bolney stating: 'It is time for me to move on and start my next challenge, knowing I have left Bolney Wine Estate and my family's legacy in safe hands.'
41 Ageing on the spent yeasts from secondary fermentation which softens and adds richness to a sparkling wine, something especially essential in English sparkling wine for taming acidity. See 'Fizz Wars', chapter 8

commercial Cava but according to Linter, the company will be 'boutique' rather than 'hugely mass market'. It wants to 'focus on premiumisation', she says, using the extra money to age the wines for longer. Bolney currently makes about 250,000 bottles a year and the plan is to take this up to half a million, of which 60 per cent will be sparkling wine, using vineyards outside of Sussex to hedge its bets in an uncertain climate.

It is very early days in the acquisition so it's still not clear exactly what this will mean for Bolney but it seems that Henkell Freixenet is not about to turn its new English winemaking operation into a budget brand for English fizz – something that would be almost impossible to do given the cost of growing grapes here. All the investment money could move Bolney into the first rank of English wine producers. The new owners will keep the site as a tourist destination – something, judging by the packed café on the day of my visit, that it does very well – but the new parent company is also well-placed to increase Bolney's presence in overseas markets. Freixenet already sells in Norway, China, Japan, America and Holland. As Lucy Auld, head of marketing at Freixenet Copestick, said in an interview with *Vineyard* magazine: 'Our reach is worldwide and we do see strong potential in the export market for Bolney wines.'

Other wine companies closer to home are also investing in English vineyards. In 2017, Boutinot Wines bought Henners vineyard in Sussex. Despite its French-sounding name, Boutinot is a wine merchant based in Manchester which has been making wine in France since 1990 and South Africa since 1994. The acquisition of Henners was its first foray on home turf.

Henners was established by ex-Formula One engineer Lawrence Warr in 2007. (Oddly, he's not the only ex-motor racing person in English wine – there's Simon Woodhead at

Stopham in West Sussex.) One shed at Henners provided a home for Warr's racing cars and the other the winery. The name came from an ancestor of Warr's, Henners Dubois, who came to England to escape the French Revolution. The first vintage was 2009 and quality was high right from the start. I remember being delighted by the 2010 vintage at a time when many English sparkling wines were painful to drink.

But perhaps because production was so tiny, at around 15,000 bottles per year, Henners never got the attention it deserved. According to Tom Whiteley from Boutinot, Warr found things 'harder than anticipated'. He added that Boutinot was looking for land in England to make wine and as the firm had been distributing Henners since 2012, acquiring the winery was a logical step. Boutinot moved out the Formula 1 cars, built a new winery and has taken production up to 100,000 bottles per year, with the capacity to make 350,000 under winemaker Collette O'Leary, who has worked in California and South Africa. There's been a rebrand, with smart new labels, and the majority of production is now non-vintage to ensure a consistent product. Boutinot has an interesting business model, and, like Henners, it flies a bit under the radar. 'We don't spend money on advertising,' says Whiteley. Instead it's a major supplier to independent wine merchants, bars and restaurants, so despite having almost no brand presence, its wines are everywhere.

Boutinot's acquisition of Henners shows how an experienced wine company can change the fortunes of an obscure English sparkling wine brand. We're likely to see more of such transformations in the coming years.

CHAPTER 8

Fizz wars

'Why would you compete with Prosecco?'

Charlie Holland

Charlie Holland, the CEO and chief winemaker at Gusbourne, says 'it could ruin this amazing thing we have', while veteran industry commentator Tom Cannavan has described it as 'dangerous'. What could this frightening thing be? Brexit, perhaps, or some kind of ferocious new vine-eating pest from Japan? Rather prosaically, they're actually referring to a method for making sparkling wine. That method is known as 'Charmat', and it's got the industry in a tizzy. If you want to get a strong reaction from someone in the English wine business, uttering this harmless-sounding word is the way to do it.

To understand exactly why this might be, we have to take a journey back to a time when Champagne didn't sparkle. For much of history, Champagne, as the nearest quality winemaking region to Paris, was the wine of choice for French kings. But by the 17th century, improvements in transport meant that other regions which could ripen grapes far more reliably, like Burgundy, were beginning to send more wine to the capital.

The wine produced in Champagne back then would have been a sort of pale rosé, known as a vin gris, produced by fermenting black and white grapes together. The problem is that Champagne is too far north to reliably ripen grapes every year. Some years there would be a good harvest but too many years the wine would be mean and lean – easy prey to producers of more opulent, deep-coloured red wines from further south. Champagne could have become a viticultural backwater, like the vineyards around the city of Paris (you might be surprised to learn that there is still a little walled vineyard in Montmartre making tiny amounts of wine). Yet thanks to a bit of 17th-century high technology, Champagne thrived as a wine region and has become a byword for prestige and luxury the world over. That technology? The wine bottle.

High-strength glass capable of transporting wine was invented in England around 1630, probably by the polymath Sir Kenelm Digby. Previously, glass was much too fragile for anything except decanters, drinking vessels and plate glass. English scientists like Digby and others then quickly saw the potential in these new, stronger bottles for a special kind of alchemy – wine with bubbles. In cold climates like Champagne, wine would stop fermenting in the autumn and then start again in the spring. Similarly, wines shipped to England would often be full of bubbles when the weather warmed up, and these slightly sparkling wines were enjoyed by the aristocracy. Sadly for them, however, there would be just a few weeks of frivolity before they stopped fermenting. But with these new bottles, the bubbles could be preserved (bottles before this point would have just exploded under the pressure). Simply fill them with wine that hasn't fully fermented so it contains sugar, then cork securely, leave for a few months and voilà! Sparkling wine.

It was then discovered that you could control the fermentation by adding sugar. Reverend John Beale gave a paper to the Royal Society on 10 December 1662 outlining a method for making sparkling cider by adding 'a walnut of sugar' to finished bottles. A walnut was around 20g, a very similar amount used in many Champagnes today. Just two days later, Christopher Merret gave a now celebrated paper about applying a similar process to wine. This technique crossed the channel in the early 18th century (the first patent for English style bottles was in 1709) and sparkling Champagne began to take off. It was something of a hit-and-miss affair but gradually, through the works of pioneering winemakers like Madame Clicquot and scientists like Louis Pasteur, the Champagne method evolved.

By the end of the 19th century, something like the modern technique had been perfected. It involved taking a dry, still wine of about 10% ABV, usually made from chardonnay, pinot noir and pinot meunier grapes. This base wine would be extremely high in acidity and not terribly pleasant to drink. It would then be blended with some older reserve wines (reserve wines being richer and softer for having been aged longer), a solution of sugar and yeast (known as a liqueur de tirage) added to it, put in a strong glass bottle and left for between eighteen months and five years. This would create the fizz, though the wine would be cloudy from the deposit of the yeast cells.

To counter this, a technique known as riddling was pioneered

113

form of flattery, but either way, the British public were prepared to pay Champagne prices for Champagne-esque wines. Not that most customers necessarily know all these details, but the assumption is that if you see a bottle of English Sparkling Wine, it's going to be a bit like Champagne.

The regulations do, however, lead to some peculiar anomalies: Peter Hall at Breaky Bottom, one of the country's greatest producers of fizz, isn't allowed to call many of his cuvées English Sparkling Wine as he uses seyval blanc rather than the classic ESW-approved varieties. And the Champagne method isn't the only game in town for making sparkling wine. Winemakers in Prosecco in Italy use the tank or 'charmat' method. Though it's most commonly associated with Prosecco, this technique was actually invented in Bordeaux by a Frenchman, Eugène Charmat. It involves creating a base wine and then adding yeast and sugar to begin a secondary fermentation in a pressurised tank (rather than a bottle, as with Champagne). The wine is then chilled, filtered to remove yeast and then bottled under pressure, usually with some sweetness left in the wine (or a dosage can be added). This method is also used to create Asti Spumante, Sekt in Germany and all kinds of other inexpensive sparkling wines. The resulting wine usually has less fizz than using the traditional method, and tends to taste of the fruitiness of the grape rather than the mature aromas like Marmite and bread that you associate with English sparkling wine or Champagne. Most importantly, rather than 18 months, it takes a matter of weeks to achieve, meaning you can get your wine to market quicker. You can see why some producers might like it.

And you can see why some don't. Simon Roberts, winemaker at Ridgeview, argues that traditional-method sparkling wine boasts a 'value added product' – if it looks like Champagne, you can charge more for it. He is worried about the industry 'going off in the wrong direction' and thinks that it needs more regulation. Charlie Holland of Gusbourne, which sells sparkling wines for up to £195 a bottle, is also not a fan of Charmat wines. 'Why would you compete with Prosecco?' he asks, adding that many people won't understand the difference between the two styles, and therefore Charmat wines risk 'damaging the name of English sparkling wine'. Meanwhile at Rathfinny, Mark Driver says that part of the reason for campaigning for a separate PDO for Sussex was to 'protect us from Charmat' which 'confuses consumers'. Driver worries that people who don't appreciate the techniques behind ESW will try a Charmat wine, find it different and then be put off all ESW for good.

In an interview with the newspaper *City AM*, Tom Cannavan commented: 'The English wine industry is still maturing and going through a period of rapid development, with new players and new wines. Some of the established producers are striving to move the market 'up', with luxury cuvees at £100 a bottle, single-vineyard wines, special aged releases . . . All of that is predicated on English sparkling wine's story of using the same method as Champagne, in a similar climate and, in some cases, similar chalk soils.' Or as Stephen Skelton puts it: 'The story that sells best is that Great Britain's sparkling wines are "as good as Champagne."'

Flint Vineyards in Norfolk seems like an unlikely base for a revolution that could undermine the English wine industry, and the lanky, softly-spoken founder Ben Witchell is an unlikely revolutionary. And yet wines like his Flint Charmat Rosé are the kind of thing that are getting everyone all hot and bothered. Wichell used to work in IT but gave it up to become a winemaker. He studied at Plumpton College and graduated

VINES IN A COLD CLIMATE

top of his class (it was his head of marketing who told me this; Witchell is far too modest to bring it up). Following some time working in Beaujolais, he and his wife set up Flint in 2016 on a comparative shoestring, in partnership with a local farmer.

As his vines matured, Witchell had to buy in all his grapes to make his first wines. The sparkling wine came about by accident when he bought a parcel of grapes in 2017 which included some rondo, a much-maligned red grape, among other varieties. 'What the hell do I do with this?' he thought to himself. His first instinct was to blend it with some other varieties and make a rosé. Then he had the idea of making it fizzy. He spoke with David Cowderoy, one of England's great wine pioneers, who runs a company called BevTech in Sussex. As well as selling equipment for the wine and beer industries, the firm offers services such as bottling and producing Charmat-method fizz. So Witchell sent his base wine to Bev-Tech, and Cowderoy fermented and bottled it.

Witchell was so pleased with the results that he wondered why people hasn't done it before. You'd never mistake the wine for a traditional method English sparkling wine but nor is it a Prosecco copy. In fact, with its big cherry sherbet flavours, it's like stepping into an old-fashioned English sweet shop. People love it and he sells out every year despite, at £24, it being about the same price as an entry-level, traditional-method sparkling wine.

Though he doesn't want to be seen as combative, Witchell is candid about why certain producers are so anti-Charmat. 'People have invested too much and have too much to lose,' he says. Producers sitting on millions of pounds worth of maturing sparkling wine don't want competition from something that can be made in hours rather than years. But he bats away the idea that Charmat will cheapen the image of English wine 'It's easy to make average English sparkling wine, anyone can do it. There's more of a danger from people making boring wine [than from Charmat].' He has a point here. A lot of English sparkling wine, like much of Champagne, is competent rather than thrilling. That might have been enough when competency was more than people expected, but expectations have risen since.

Witchell thinks that Charmat can actually *help* the quality of classic sparkling wines. The very best traditional-method wines only use the gentlest pressings of the grapes. Witchell makes a tiny quantity of special multi-vintage English sparkling wine from the first pressing, known as the coeur de cuvée, with later pressings going into his Charmat wines. 'Charmat can augment quality by using less fine wines,' he said. He thinks the more aromatic Germanic varieties like reichensteiner and bacchus are particularly suited to the method.

Making only 35–50,000 bottles a year, Witchell isn't a threat to the likes of Gusbourne and Ridgeview, and he seems popular within the industry. The same can't necessarily be said of Monaco-based businessman Mark Dixon. Dixon made his fortune from Regus serviced offices. The wine side of his business is called MDCV and includes Château de Berne, apparently the second largest producer of Provence rosé. In 2017 he moved into English wine with the purchase of Kingscote Estate in Sussex, to be followed by that of Sedlescombe.

Both of these are now up for sale again but Dixon has no intention of moving out of the English wine industry. Far from it. Since 2017 he has planted an astonishing 17 million vines near Gravesend in Kent, which he is farming organically.

The aim is to produce 5 million bottles of wine a year by 2025. To put that in perspective, in 2018, a bumper year, the entire English wine industry produced only 15 million bottles.

As you might expect, Stephen Skelton has some forthright things to say about Dixon, among them a description of him as 'a rich twat'. He's also not convinced that Dixon will be able to achieve his ambitions, claiming that the vineyard sits in a 'frost pocket'. But it's not just Dixon's ambitions that have ruffled the feathers of the English wine industry. The name of his Charmat wine – Harlot – led to it being described as England's 'most controversial wine'.[42] It's sometimes hard to know whether people are really offended by this name or just performing. After all, Fiona Beckett, wine writer for the *Guardian*, had a minor spasm at the sight of the Union Jack on a bottle of Morrisons English Sparkling Wine, and you should have seen the outrage on social media when the English Vineyard Association merged with the English Wine Producers Association to form Wine GB, complete with Union Jack branding. Nevertheless, using a term for a promiscuous woman to brand a wine did generate some genuine outrage. Adrian Pike of the producer Westwell described it as 'a teenager's wet dream' and blackballed Dixon from membership of the association of Kentish wine growers, the Wine Garden of England.

Dixon declined to be interviewed for this book but I did speak to his head of marketing, Emma Clark, a veteran of the wine trade after stints at Oddbins and Enotria. She doesn't see the problem: 'Unless you are 70-plus I don't know anyone who would use "harlot" as a derogatory insult,' she says. Apparently the name is 'all about reclaiming the word', she says, to reference

---

42 *City AM*

'strength of character, determination, not having boundaries, not being pigeon-holed, not being an old person,' rather than as a term for a prostitute. It's also aimed 'at a younger demographic' than those who are offended, or claiming to be offended. Whether you are convinced or not, Harlot upset all the right people and made itself the most talked-about wine in England. Job done. Though Ben Witchell from Flint is more relaxed about the whole thing: 'A wine called Harlot? Good luck to him.'

The idea for the wine came through conversations with advertising agency JKR. 'Rather than making a style and selling it, we would come up with what the consumer wanted,' Clark says. And what the consumer wanted, apparently, was Prosecco. By the same token, 'We want a slice of the Prosecco market,' admits Clark. But there has been a certain amount of scepticism among the trade that consumers will be happy to pay around double the price of most Proseccos for an English Prosecco-style wine. Clark argues that people are happy to pay more for a locally-made wine, especially with the brand's organic credentials (though it doesn't as yet have organic accreditation). Yet rather than emphasise the wine's sustainability, Harlot's marketing majors on a rather humdrum rebelliousness featuring extravagantly dressed people of indeterminate sex who look like they have stepped out of a 1990s nightclub.

The name might be a bit of a joke but the wines, especially the rosé, are actually very tasty – a definite cut above most Prosecco. Harlot is a blend of the three Champagne varieties along with bacchus, with the fizz coming via the Charmat method. It is made by a serious winemaker in Jérôme Barret, a Frenchman who has made wines in Champagne as well as less-conventional wine-producing countries like Russia, China, Japan and Armenia. He has also lectured at the Champagne

121

Institute of Oenology, so it's safe to say he knows a bit about fizz. Yet he's making Charmat method wines in England. Again Skelton is sceptical. 'Fucking French consultants,' he says. 'What you do in Burgundy and Champagne doesn't work here.'

Clark thinks that the two markets are so different that Charmat wines are not going to affect English Sparkling Wine (ESW). 'We can grow excellent grapes, we have the technology for Charmat style, so why not produce the English version? It's complementary to the traditional style.' MDCV also produces ESW but Harlot is aimed at people who might find such wines too expensive, and sits in the gap between Prosecco and ESW. Clark compares it to the relationship between Prosecco and Franciacorta, a Champagne-style fizz also made in Northern Italy; one doesn't detract from the other, they have different markets and are enjoyed on different occasions.

Part of the problem in England is that there is no generic name for English Sparkling Wine that resonates with customers. Champagne is Champagne, Prosecco is Prosecco, and you know roughly what you're getting. There have been various attempts to invent a name for a traditional method English wine. Coates & Seely came up with 'Britagne' but according to owner Christian Seely 'nobody adopted it and we thought perhaps it's not such a great idea so we discreetly dropped it'. The firm, however, does have the words 'méthode Britannique' on the bottle, a play on 'méthode Champenoise'. I proposed that you could call ESW 'Digby' after the inventor of the wine bottle but the name had already been taken by a sparkling wine brand. Wine writer Anthony Rose suggested FRED, F for fermentation, RE for reumage and D for disgorgement, which sounds fun. I can just see people calling for 'Another bottle of your finest Fred, waiter!' It's certainly better than an attempt by former *Guardian* wine writer Malcolm Gluck to make it a 'Pippa' after Pippa Middleton, Prince William's sister-in-law.

Such branding is very much still a work in progress. According to Brad Greatrix of Nyetimber: 'There is a consultation process underway within the industry because it is clear to all that the PDO in its current form does nothing to serve consumers or producers.' Wine GB wants to make sure that the production method is clear on the label by use of the term 'classic method'. If only producers could just call it 'English Champagne' and be done with it, but apparently the French wouldn't like that.

There are now quite a few Charmat-method wines on the market and most I have tried have been good quality if relatively expensive for such simple wines. The pressurised tanks needed to make these wines are extremely costly, so most brands make use of Cowderoy's BevTech facility. But there are cheaper ways of injecting fizz. Over at Chapel Down in Kent, they produce something even more shocking to the traditionalists – carbonated bacchus. No secondary fermentation needed, just put bubbles in it using a giant Soda Stream. It's actually a nice wine as long as you're not expecting something like Champagne. Chapel Down winemaker Josh Donaghay-Spire says that it preserves the 'bacchusness of bacchus' – something that would be lost using the traditional method.

Many producers looking for cheaper alternatives, however, are going back to sparkling wine's 17th-century roots with something known as *petillant naturel*, or 'pet nat' for short. This technique involves bottling the wine while it is still fermenting. The sugar left in the wine keeps it fermenting in the bottle and the resulting carbon dioxide is absorbed into the wine. It's a style that's become very fashionable in hipster bars in Paris, New York and London.

Will Davenport makes one on the advice of his distributor, Doug Wregg at distributor Les Caves de Pyrene, who he describes as his 'pet nat mentor'. According to Davenport, pet nat wines can be a hit-and-miss affair: 'You have to measure the sugar level assiduously and then bottle at the right time. It works with a slow ferment but one year the ferment was so fast that the last-bottled wines had much less sugar in, so they had barely any fizz. Some years we get it right, some years we don't make it.' He makes the wines because they sell, even if personally he isn't a huge fan.

The technique produces a relatively simple-tasting cloudy wine with less fizz than a traditional English Sparkling Wine. It's handy for producers' cash flow as the wine can be released soon after harvest and, unlike Charmat or Champagne-method, doesn't require any expensive equipment. You can see why some winemakers love them – Westwell's Adrian Pike describes them as 'an instant expression of terroir'. The seeming ease with which you can make a pet nat has inspired some amateur winemakers to chance their arm, often via colourful labels and silly names like Fizzy Bum Bum, made by Bin Two wine shop in Padstow.

Master of Wine Tim Wildman put on a tasting of English pet nats – or 'Brit nats' as some people are calling them – in London in October 2022. According to one attendee who wished to remain nameless, 'Most of the wines fell somewhere between charmless and appalling.' A problem inherent to the technique is that it's very difficult to control the fizz of the wines, meaning that sometimes, on opening, most of the contents can jump out of the bottle onto the floor. As another attendee at the tasting, Sophie Thorpe, wrote in *Club Oenologique*: 'The biggest problem was stability, with a handful of wines "gushing", sometimes uncontrollably, rendering them commercially questionable.' But she was more positive about the wines overall: 'The quality surpassed expectations, though – both for me and almost everyone in attendance.'

Controversy persists, however, around the concept of people 'cheating' in their winemaking, with Pike muttering darkly about 'pet nats that aren't'. To make a true pet nat, the process needs to be undertaken during the harvest, taking up valuable time when the winemaker could be processing grapes. As a result, some contract winemakers in England offer a pet nat service which involves taking a fully-fermented base wine and adding sugar and yeast, to make a sort of hybrid between pet nat and Champagne-method. This is similar to a style of sparkling wine known as Col Fondo which was how Prosecco used to be made before the region adopted the Charmat method. Hugo Stewart at Domaine Hugo in Wiltshire describes it as 'the missing link between pet nat and English Sparkling Wine'. To make his, he adds the still-fermenting wine from one vintage to the finished wine of another, and then bottles it under a crown cap like you would find on a beer bottle. It produces a wine that's much less fizzy than classic sparkling wine, with the yeast remaining in the bottle and lending a slightly cloudy appearance.

Even more daring is near neighbour Tommy Grimshaw, winemaker at Langham, who produces a sparkling wine in kegs for local restaurants. Unlikely as it may sound, Grimshaw was inspired by a visit to the Wetherspoons pub in Dorchester, where he saw Prosecco come out of a tap and wondered if he could do something similar. He spoke to Key Keg, a company which sells kegs to the brewing industry and advised him against the idea. Nevertheless, he ordered three kegs to undertake an experimental trial. His approach involves adding a little sugar and yeast to an already fermented wine, putting it in a keg and letting it re-ferment to produce a slightly sparkling wine. To the usual Champagne grape mixture, Grimshaw brought in some ortega and madeleine angevine grapes to make the wine more aromatic. It's proved a massive hit with local venues and worked so well that he immediately bought another 18 kegs. Grimshaw thinks the wine offers much better value than a bottled Charmat wine as well as being better for the environment.

But does the consumer actually give any thought to what the wine comes out of or how it is made? Does he or she even care? Ben Witchell says that none of his customers – and he sells a huge amount cellar door – are particularly interested in the production method of the sparkling wine. They like the taste and they are prepared to spend nearly £25 a bottle on it. 'And if your wine is of such good quality, why worry?' Why indeed.

CHAPTER 9

Big wine

'The original brief was to make a Provence-style rosé – until I pointed out that grenache and cinsault don't grow in England . . .'

Henry Sugden

If you visit a country fair in England, you'll almost always come across a couple, usually of a certain age, selling their own wine. These bottles won't be the amateurish affairs of yore but often slickly packaged and well-made sparkling wines that proudly tout their provenance. But scratch the surface and all is not what it seems. While there are something like 500 vine growers in England, there are only around 100 actual producers. The vast majority of English wines are produced by contract winemakers.

At a tasting put on by the Vineyards of Hampshire organisation, most of the wines in the room were made at just two wineries in that county – Hambledon and Hattingley Valley. It's a similar story all over the country. If you drink a Welsh or West Country wine, it's more than likely to be made at Three Choirs in Gloucestershire. Meanwhile in Staffordshire, Halfpenny (pronounced ha'penny) Green provides a similar service for growers in the Midlands. Not that there's anything wrong with this; all these wineries make wine to a high standard. But it can be a little misleading to the consumer who probably thinks the whole process, from grape to bottle, is conducted in-house.

Contract winemaking is such a growth area that in 2019 Henry Sugden, one of those men who prove the old adage that you can take the man out of the army but you can't take the army out of the man, opened Defined Wines near Canterbury with the express purpose of focusing solely on contract wines. Defined Wines does not own any vineyards or have its own label. Instead, it works with everyone from tiny husband-and-wife producers to some of the biggest names in the business. (On my visit, I saw wines being labelled up for a very big brand, but was sworn to secrecy as to its identity.) The team is headed up by Nick Lane, who used to make wine at Cloudy Bay in New Zealand – part of the LVMH group that owns Moët et Chandon among others. When I visited, the facility was gearing up for its third harvest, the tricky 2021 vintage. It was September, wet and overcast, and expectations were modest as to the quality of the grapes likely to be coming in.

As is often the case when asking a winemaker about their wines (and not just in England), Sugden began with a swipe at the competition. He isn't impressed with the overall standard of English wines: 'There's quite a lot of home brew around. I had an award-winning rosé that was oxidised. If you go to an artisan winemaker in France, it will be good. Not so in

England.' He describes the industry as 'a bit Mickey Mouse compared with the rest of the world.' Defined Wines struggled to find the right home-grown winemaker and didn't want to poach an established one so brought Lane over from New Zealand. Despite this, Sugden thinks the industry is very well placed to grow: 'It's not Venezuela – this is the centre of international wine trading. There are more Masters of Wine here than in any other country. This [growth] is happening in a sophisticated wine-drinking culture.'

For growers, the advantage of working with a contract winemaker like Defined is access to the very best equipment. A Coquard press from France, the kind of thing used by almost all sparkling winemakers in England, will cost between £40–60,000 depending on size. Simon Roberts from Ridgeview said in an interview with Vineyard magazine:[43] 'Without the contract winemaking we simply wouldn't be able to justify equipment like this. Dad always used to joke that it was his Aston Martin. I love this press and the quality of juice it produces not only goes into our wines, but our contract customers too, and that in turn benefits the wider industry.'

Defined also has a state-of-the-art laboratory so that it can monitor every step of the winemaking process. It was manned, when I arrived, by Poppy Seeley, who had worked in California, South Africa and New Zealand, after studying at Plumpton College. She wasn't entirely complimentary about the education she received there: 'They didn't have chemistry equipment, didn't have a proper lab. I was frustrated at Plumpton.' She thought many 'Plumpton graduates think they can walk into a job' whereas she 'learned far more *on* the job'.

---

43 *Vineyard* magazine

Defined opened full time in 2020 and its current production is around 70 per cent sparkling and 30 per cent still wine. 'Some clients know what they want, some don't,' Sugden told me. He talked about one brand whose 'original brief was to make a Provence-style rosé until I had to point out that grenache and cinsault [the classic Provence rosé grape varieties] don't grow in England.' If a client wants something like Chapel Down's Flint Dry – probably England's best-selling still wine – then he's happy to make that. The team has to deal with all kinds of levels of experience which sometimes means growers who bring in less than excellent fruit. 'We have people who say "work your magic" on bad grapes,' Sugden said. As a result, he gets involved with viticulture and marketing too, and says his team visits every vineyard it works with. 'We don't want to make 30,000 bottles that don't sell.'

Ultimately, though, winemakers are at the mercy of the grapes they are given. Wine writer and sometime winemaker Tom Hewson says: 'I was going around English producers tasting base wines, and one winemaker was very honest with me about what sort of grapes they had received, and gave me a taste of something made under contract. I tasted it and it just wasn't very ripe.' Hewson thinks that the contract wine model doesn't incentivise quality. 'They [contract wineries] want to get the fruit in, as they get paid by the bottle. They don't have a clear-cut benefit in making sure ripeness is really high.' He thinks that many grapes that wouldn't be deemed of good enough quality to go into the top brands still make it into contract-made wines. 'I know of another vineyard that only uses the top third of production of its own wines and sells off the other two thirds,' he adds.

On my visit to Defined, I met Nayan Gowda. Born in Derbyshire and from an Indian immigrant family, Gowda studied winemaking at the University of Adelaide. He has had an incredibly varied career which spans making huge-volumes wines like Black Tower in Germany, working in the wild East in places like Kazakhstan and Ukraine, and helping small-scale farmers with the most rudimentary equipment in Bolivia. He described the Defined team as 'people working at the top of their game'. This is the first time he has made wine in England and I asked him what was the main difference with some of the other places he's been. 'Everything works!' he said.

According to Gowda, Defined has three sorts of clients. The first are those who say roughly what style of wine they want to make, and let Defined do the rest. Certain English wines are pure brands in that they own neither a vineyard or a winery, and don't actually make the wine. These include Folc, a rosé brand owned by entrepreneurial couple Elisha Rai and Tom Cannon, whose expertise is in marketing and branding and who wisely leave the sourcing of grapes and the winemaking up to the team at Defined. These, says Gowda, are the easiest clients to deal with.

Then there are tiny growers, those who produce under a tonne of grapes a year. This is too small a quantity for their wine to be treated separately; according to Gowda, the minimum amount to qualify for bespoke treatment is two tonnes. So these growers will have their grapes mixed in with fruit from similar growers to produce a consistent wine, similar to the co-op model on the Continent. It means that the wine from the little husband-and-wife producer you see at the county fair might well be at least partly from grapes that they didn't grow. It's a little more transparent at Halfpenny Green[44] in Staffordshire, which was

---

44 *The Vineyards of Britain*, Ed Dallimore

founded by Martin Green in 1985 – ancient history in English wine terms – and works with around 30 cooperatives (groups of growers) whose members pool their grapes and get back the same wine, in addition to 40 separate, larger growers who get their own individual wines.

Finally, there are those who micro-manage every step of the process, from the pressing of the grapes to fermentation temperatures and bottling. Defined have one huge client like this, the one I'm not allowed to mention. Simon Roberts from Ridgeview, which has similar clients, told me about the logistical requirements involved in supplying Waitrose and others with own-label wine: 'You can't work with big customers without accreditation from an organisation like SALSA [a UK approval scheme that helps local and regional food and drink producers supply their products to national and regional buyers]. So you have to do things like cleanliness to a certain standard. We wouldn't have the clients we have now without it.'

Over at Hattingley in Hampshire, its big client is none other than Pommery. This prestigious Champagne house is currently building its own winery, also in Hampshire, but in the meantime, all the wines are made at Hattingley – to very strict parameters, according to winemaker James Rouse. The Hattingley set up is impressive. It's built on the site of an old chicken farm, and despite the enormous capacity, is set in a series of brick- and wood-fronted sheds that fit beautifully into the landscape. In the warehouse, one side is devoted to maturing Hattingley own-label wines, and the other to contract wines. The whole enterprise is set up for handling large quantities of grapes from multiple sources. You imagine that the hard-pressed team at Chapel Down would kill for such facilities.

On my visit, Rouse was adjusting to life without Emma Rice

who'd been with Hattingley since it was founded in 2008 (she was replaced as head winemaker by Rob MacCulloch MW). The phones Rice was clearing out of her desk were testament to her length of service, from Nokia bricks to Blackberries to iPhones. She described some clients as more 'high maintenance' than others. 'We got rid of the difficult ones,' she said, before recounting all manner of enquiries over the years, from new growers asking how to pick grapes to calls during the harvest offering her their entire crop. She's even had people with a vine in their garden getting in contact to see if she wanted the grapes.

Rouse relishes working in a cool climate. I couldn't place his accent at first, and thought he might be Irish, but he turned out to be Canadian. After working for Nyetimber, followed by many years in the Australian wine industry, he decided to settle in England, where he has family. There are notable differences in approach, he says. 'In Australia you have to spend a lot of time fining the juice [removing large particles] and adding tartaric acid. Here the fruit is pristine – you don't have to add anything.' Nevertheless, most large-scale sparkling-wine producers such as Hattingley and Defined do incorporate certain elements to make their wines more consistent and to speed up the process. This will start at the press stage when the grapes come in from the vineyard. It's standard practice to add sulphur dioxide to prevent the juice oxidising and going brown, and to stop wild yeasts starting fermentation. But every winery has its own techniques and makes wine in its own style. Just down the road from Hattingley at Hambledon, they aren't concerned about keeping the oxygen from the juice. Winemaker's assistant Sam Picton argues that many of the textbooks that people use are from California or Australia, which are home to warmer temperatures and less acidity in the grapes. By comparison, he thinks English grapes can take a bit of oxygen, which makes for a more stable wine further down the line.

Poppy Seeley from Defined describes the main challenge with making wine in England as the opposite to what she encountered in South Africa and Australia. There it was about lack of acidity, whereas in England it's all about tempering the high acidity. In the past it would have been routine to de-acidify either the wine or the grape must (juice) before fermentation. This isn't so common now as, through better vineyard management, understanding of varieties and site, and, of course, global warming, grapes are coming in riper. According to Rouse, they avoid it at Hattingley as de-acidifying is 'a brutal way to treat wine'. But the practice does still happen, via a technique known as ion exchange or by adding chemicals such as calcium carbonate (chalk), potassium carbonate and potassium bicarbonate, either before or after fermentation. In his 1977 book *Vineyards in England and Wales*, George Ordish described how 'calcium carbonate leaves a calcium taste in wine' – which is why some English wines used to taste chalky. The other technique is to water the grape juice down and then add sugar to bring up the alcohol levels – though this will leave you with a severely diluted wine.

Adding sugar, in contrast to de-acidifying wine or grape juice, is routine. Almost all English wines will turn to a process known as 'chapitalisation' to boost alcohol levels, and you'll see huge sacks of Tate and Lyle at most wineries. PDO (Protected Designation of Origin) rules for making an English Quality Sparkling Wine allow for the addition of sugar to create 3% worth of alcohol. But in cooler years such as 2021, where less ripe grapes will come in with lower levels of potential alcohol, there's dispensation to go up to 3.5%. It's worth considering what this means in practice: you could start with grapes picked at 7% potential alcohol being bumped up to 10.5% using cane sugar, and then, in the secondary fermentation process, gaining another 1.5% ABV from sugar.[45] So in a 12% sparkling wine from a cool year, around two-fifths of the alcohol could come from almost tasteless cane sugar rather than the grape. Tom Hewson reckons that 'international winemakers would probably class more than 50 per cent of the fruit grown across the country as underripe. In vintages like 2019 and 2021, the figure would be higher.'

The more quality conscious producers, however, don't go to these extremes. Tommy Grimshaw at Langham in Dorset says he never adds more than 1.5% ABV from sugar. Grimshaw also doesn't use enzymes or cultured yeasts for the first fermentation, which is very unusual. In conventional winemaking, the winemaker will add things at every step of the process to prevent oxidation, speed up clarification, aid fermentation and to create a consistent, stable wine. I should add here that there's absolutely nothing wrong with these processes. Some of the best wines in England and indeed Champagne are made using these techniques. If you want to make wine on a vast scale then you have to use them.

Once the wine has finished fermenting, the bottom of the vessel will be full of spent yeast cells known as lees. These are allowed to settle before the wine is pumped off and then it's usual to remove any remaining solids from the wine using a fine clay called bentonite. 'We use a lot of it,' says Gowda. 'It's a finite resource and creates problems downstream. But people want very clear wines.'

---

45 See 'Fizz Wars', chapter 8

Wine that is being made into sparkling will then have sugar and yeast (and sometimes enzymes to help with fermentation) added before bottling. The re-fermentation in the bottle will produce the bubbles. After the spent yeast has been removed in a process known as disgorgement, the wine is topped up, usually with a mixture of sugar and wine, though this could be grape juice or just plain wine, before being sealed with a cork. There are some producers like Peter Hall at Breaky Bottom who do nearly everything on site but lack the equipment to do this last part of the process reliably, so contract it out to the likes of Ridgeview or Wiston.

Making an English Sparkling Wine is as much about the winemaking process as what goes on in the vineyard, which means that the winemaker's stamp is all important. It's possible to make a fair guess at the derivation of the wine just from tasting it. At Hattingley, Emma Rice's technique involved fermenting a portion of the wines in oak to produce a broader, richer style of wine. Perhaps the winemaker with the most distinctive style is a twinkle-eyed Irishman called Dermot Sugrue, who was until recently head winemaker at Wiston Estate in West Sussex. Sugrue has had one of the more interesting careers in English wine. He started at Nyetimber during the Andy Hill years of the early 2000s before moving on shortly after Eric Heerema took over, in 2006, to join Wiston. Sugrue's big thing is acidity. He loves a high-acid wine and whereas most English winemakers rely on malolactic fermentation to reduce acidity, Sugrue actively avoids it. Malolactic fermentation is a process whereby bacteria turn the sharp malic acid (like that of a green apple) into lactic acid (like you find in milk). It normally happens naturally a few months after fermentation but it can be encouraged by adding bacteria, or blocked by adding sulphur dioxide. In her second vintage after taking over the winemaking at Nyetimber, Cherie Spriggs returned to using malolactic fermentation and says she's much happier with the wines.

Sugrue's wines are among the best in the country, though some can be a bit too high in acidity for my taste. His wines have won many awards not just for Wiston but for clients like Digby and Ashling Park, which was proudly displaying its President's Trophy from the Wine GB awards when I visited. Gail Gardner, Ashling Park's managing director, clearly loves working with Sugrue and seemed a bit sad that he was leaving.

My visit to Wiston coincided with Sugrue parting ways with the winery in order to concentrate purely on his own-label wines, Sugrue South Downs, with his new wife, Ana Đogic. I was warned that he might not be on top form as he had just come back from his stag weekend. Sure enough, when Sugrue lumbered into sight he was covered in cuts and bruises, though whether this was the consequence of over-indulgence or mountain biking on the South Downs – or both – wasn't entirely clear.

Also in attendance was Kirsty Goring from the family that owns Wiston, and there seemed to be a degree of tension between the pair as they interrupted and competed with each other to explain the wines. Such was the level of bickering at one point that, had this been an eighties sitcom, they would have started kissing. Joking aside, it's clearly a complicated relationship, with equal parts affection and exasperation on both sides. There's evidently a contradiction between the Goring family's understandable desire to have their wines put first, and the success of Sugrue South Downs. This blew up at one point when, after trying the Wiston range, it was time to taste Sugrue's own wines, which are made at Wiston from various smaller vineyards. While we were tasting the aptly named 'Trouble with Dreams,' Goring, perhaps worried that her wines were being upstaged, interrupted Sugrue to open one of Wiston's rarer and more expensive wines, a 2015 Blanc de Blancs, which was indeed superb.

Sugrue described contract winemaking as 'a young man's game' and he was clearly relishing the forthcoming freedom that would come through making his own wines. Emma Rice from Hattingley, who is moving on at the same time, echoed Sugrue's views. Sugden at Defined told me that during harvest, the winery runs for 24 hours, while even at smaller wineries like Langham, Grimshaw and his team sleep in a caravan on-site, often working until 2am before rising at 6am to do it all again. 'By the end we're just running on beer and cake,' he says.

Goring seemed to suggest that Sugrue was still going to be involved in some way at Wiston whereas he said that he was planning to build his own winery and generally looked a bit uncomfortable about the situation. You can see why someone with Sugrue's restless creativity might ultimately find working for one of the bigger wineries and doing contract winemaking a frustrating business. As we'll see, smaller producers often do things a little bit differently . . .

CHAPTER 10

Small wine

'Received wisdom can fuck off!'

Ben Walgate

Tillingham was not what I expected. The East Sussex producer is England's best-known 'natural wine' producer. I'll come to the vexed question of what exactly is meant by 'natural wine' shortly. I'll also take it out of quotation marks. Tillingham was founded by Ben Walgate, who makes wines that taste very different to the straight-laced English norm. He uses rather alternative techniques such as fermenting in Georgian qvevri – giant clay jars like the ancient Greeks would have used – or ageing his wines under flor – a layer of yeast, like they do to make sherry.

As a result, I was expecting something a little rustic. In reality, Tillingham is like a south-coast outpost of Soho House. There were young, glamorous women everywhere. The gardener

139

looked like she had stepped out of the pages of *Vogue*. Inside there's polished concrete, Brazilian music, sourdough pizza, a proper restaurant and £22 candles on sale in the giftshop. A room at Tillingham will cost around £200 a night, with *The Sunday Times* describing it as 'a magnet for A-listers as well as hipsters escaping the city'. It reminded me of a visit to Google HQ in London where there was no evidence of anyone over the age of 35; a friend said it was like walking into a cult.

I sat at the bar and drank my drip coffee until the master of ceremonies himself turned up. Startlingly blonde and blue-eyed, Walgate is one of the most recognisable people in English wine. He's originally from Grimsby in North Lincolnshire but has been in the south for a long time now, working first for an organic wine shop before going to the Isle of Wight to make Germanic-style wines about which he was non too enthused. Apparently 'the vineyard was a mess and the winery a shithole,' but he enjoyed driving tractors, fixing things himself and making wine.

In 2013, following his stint on the Isle of Wight, Walgate joined a business consortium that included his stepfather Paul Bentham and Lord Ashcroft and bought Gusbourne, the then boutique Kent winery, where he became CEO and winemaker. The Gusbourne site quickly grew from 20 to over 300 acres and the job became increasingly corporate. 'I didn't realise how big it would get,' Walgate said. 'I got further and further from the good of the harvest. I loved doing everything myself. When things went wrong I fixed them.' Disillusioned, he left to set up Tillingham.

Looking at old pictures of Walgate, clean shaven, cropped hair and wearing a tie, it's hard to reconcile with the leonine, Viking-like figure before me today. As Walgate showed me around his little empire – he employs 45 people at Tillingham – he was constantly asking questions, giving orders and even at one point indulging in a little impromptu carpentry. 'There are a million and one things that can go wrong,' he said. During my visit, his phone went constantly, reminding him, I imagine, that he had better things to do than talk to a journalist.

Tillingham came about from a partnership with landowners Lord and Lady Devonport. The winery is based in what had, since the 14th century, been an old hop farm not far from Walgate's old base at Gusbourne. Walgate brought to the table a business plan and £25,000 while Devenport put up the majority of funds by mortgaging his land, with an EU grant providing a third of the total – £70,000. Despite the establishment's glitzy appearance, Walgate insists that it's all been done on the cheap, with the most expensive outlay being a £40,000 German wine press. Both Walgate and the Devonports were passionate about working sustainably, and the Tillingham vineyards are farmed biodynamically (something we'll look at shortly) though Walgate does buy in grapes from conventionally-farmed vineyards too. [46]

As well as working in a more holistic way with the land, Walgate wanted to make wines that were less conventional than the ones produced at Gusbourne, of which he nevertheless remains a huge fan. But initially, because of a non-compete clause in his contract which prevented him from carrying out winemaking for a period of time after leaving the company, he

---

[46] As this book went to press, I learned that Walgate had stepped down as director at Tillingham. It seems the winery will continue. In a statement on Instagram, Walgate said that he 'will continue in the wings offering guidance as we transition to a new era for the business' but he is currently working on 'a new eponymous wine project (watch this space)'

made ciders, buying in juice and fermenting it in qvevri (earthenware vessels). These he sold under the Starvecrow label. I remembered tasting them back in 2018 and being extremely impressed. He still makes them and when I visited he let me try one made from Somerset apples which, though it was still fermenting in bottle, was exceptionally good. He made up for the poor 2021 harvest by making more cider – hedging for bad vintages with cider is practised by a number of other English producers.

Walgate's first vines went in the ground in 2018, so the first vintages were made solely from bought-in fruit. Tillingham describes itself on its website as a 'natural wine' producer so I think it's probably time for a definition. The natural wine movement emerged in Beaujolais in the 1990s as a reaction to the industrial winemaking techniques[47] that had taken over the region. The idea was to add as little as possible to the grapes and let the local yeasts that live on the skins ferment the wines. So no cultured yeast, no added sugar, no added enzymes and little or no sulphur. This last one is the hardest. Sulphur dioxide is anti-bacterial and prevents oxidation, and has been used in winemaking since Roman times. Making wine without it or just using a tiny amount requires perfect grapes, a spotless winery and a lot of skill. Natural wine also implies that the grapes will be grown in an organic or near organic way, something we'll look at in more detail in a later chapter. But there are no legal requirements or regulatory bodies for natural wine so anyone can say that their wine is natural. The term annoys makers of more conventional wines because it implies that there's something unnatural about their wines.

---

47 See 'Big Wine', chapter 9

From its roots in France, this natural wine movement spread all over the world, but it's particularly strong in Europe. At its best, natural winemaking is simply good winemaking, letting the quality of the grapes shine. But it has also broadened the concept of what is and isn't acceptable in a finished wine. As many producers are trying to work without sulphur, you might detect bacterial infections, volatile acidity (vinegar) or oxidation. In small quantities these can be interesting, enjoyable even, but in higher doses they can be offputting to all but hardcore fans.

Making natural wine in England has its ups and downs. Unlike most English winemaking, where the wine is chapitalised (the process of adding sugar before fermentation to boost the alcohol content), natural winemaking dictates working without sugar, meaning lower alcohol levels. And since England is normally cool at harvest time and isn't prey to the rapid effects of oxygen (which spoil the wine) that you might get in say, southern Spain or Australia, it's easier to work with less sulphur. The cold also helps clarify the wine so you don't need to filter.

Dan Ham, who made sparkling wine at Langham in Dorset before setting up the aptly-named Offbeat wines in Wiltshire, says that, 'England is good for natural wines because it's cold. 15°C in the winery is perfect for fermentation.' Ham persuaded owner Justin Langham to move to wild yeast fermentation in 2017. 'He thought I'd lost my mind,' says Ham. 'He was very nervous about it.' Especially as apart from checking and tasting the fermenting barrels 'we did nothing in the winery until August. Just waited.'

Tommy Grimshaw took over from Ham at Langham in 2020 and has continued with the natural philosophy. According to Grimshaw, with cultured yeasts, fermentation takes between six and eight weeks. So with a harvest in October and November, the wine should be done by January and then you can bottle in March. But with a wild ferment, the wine is still fermenting in April. 'There's a reason why winemakers use cultured yeast,' says Grimshaw. 'Not doing so adds stress.' If you're a commercial winery that has to clear thousands of litres of tank space for the next harvest, you don't want to be bottling in August like they do at Langham.

There's another advantage to a slow fermentation, says Ham – you can use less sulphur. He doesn't use any, arguing that constantly fermenting wine is protected by the carbon dioxide it gives off. In Ham's eyes, everything you add, such as enzymes to settle the wine, cultured yeast to ferment it and sulphur dioxide to protect it, saves time in the winemaking process, something that is vital in a large commercial winery under pressure to get wine to market and prepare for the next harvest. But if you're a small operation, less dependent on obligations to third parties such as supermarkets, then you can decide when the wine is ready.

While Grimshaw makes mainstream sparkling wines at Langham, Ham's output at Offbeat, as the name suggests, is more alternative. Some I wasn't so keen on but there's real innovation on show in the winery. Ham let me try a sample of a pinot gris that had been left in a glass demijohn outside, where it gets baked in the summer and almost frozen in the winter to produce an oxidised style of wine known as 'rancio' – literally 'rancid' in Catalan. This style of wine is usually fortified with brandy but Ham's 12% natural pinot gris had not only withstood the weather but emerged full of fruit with the trademark nuttiness that you get in such wines. Truly a very exciting wine.

Not all experiments are so successful. Natural wine lovers tend to accept a bit of inconsistency in their wines in a way that your average punter might not. This was aptly demonstrated back at Tillingham by something Walgate let me try out of a cask. He makes a sherry-style chardonnay, where the wine is protected by a layer of yeast, known as flor, floating on top while it ages. The 2020 was fresh and nutty, but sadly the 2019 vintage had developed a bacterial infection which produces a taste known in the business as 'mousiness'. It smells fine but it catches on the back of the throat with a flavour of dirty rodent cages. Not nice at all. Walgate is going to add some sulphur which might knock out the infection but might not. Such is the high-wire world of natural winemaking. It's not uncommon to detect this fault in finished wines at natural wine fairs and some customers don't seem to mind it. At Charlie Herring[48] wines in Hampshire, Tim Phillips says that the only way to work with minimum (he adds a little sulphur dioxide on bottling his still wines) additions in the winery is to be spotlessly clean. He makes sure that all the pressing and cleaning is done outside the winery so that it's completely dry inside. It's the only way to stop bacterial or fungal infections.

Tillingham is best known, however, for its qvevri wines. These giant clay pots are sunk into gravel in the ground, into which you add all the crushed grapes and simply let the wine ferment. It's a style of winemaking that has been done for thousands of years and has lasted into the modern age in Georgia, Portugal and Armenia. Qvevri is the Georgian word for these amphorae. Despite – or perhaps because of – the simple

---

48 There is no actual Charlie Herring. The name comes from Phillip's father who used to sign cartoons 'Charlie Herring' and the name stuck

approach, this form of winemaking can be challenging. 'Qvevri are an arse,' says Walgate, adding that it's possible for the wines to pick up 'qvevri taint', which spoils the wines. Phillips won't work with qvevri for this very reason. Once the ferment is done in clay, Walgate pumps them into a stainless-steel tank rather than leaving them in the ground, as would be normal in Georgia. Apparently the qvevri 'adds energy to the wines, and ages [them] a bit quicker compared with stainless steel'. Ham uses terracotta pots in a similar fashion, which he says add an earthiness and freshness to the wines.

Tillingham has become a magnet for inquisitive winemakers from all over the world. On my visit I met a German/ Italian couple who had answered an advert on Facebook to come and work at a winery. According to Walgate: 'Everyone [making wine at Tillingham] is buggering about with their own projects – perry and cider and pinot gris.' The vineyards are planted with a huge range of varieties, from the classic French ones like pinot gris, chardonnay and pinot noir to less sexy German ones like ortega and madeleine angevine. Walgate says of his wines: 'It doesn't matter what's in them so long as they taste nice. No one knows what English wine is anyway.'

Walgate produces a bewildering array of wines including skin contact wines (whites fermented on grape skins like a red wine); qvevri-aged wines; traditional-method sparkling wines; pet nats (an old-fashioned form of sparkling wine that has recently become very fashionable); single varietal wines; and wines made from a hodgepodge blend. Then there are the ciders and hybrid apple and grape wines. Tillingham seems to divide wine lovers. Some people I spoke to see Walgates's work as the most exciting thing in English wine, whereas for others the wines are over-hyped and over-expensive. They're certainly not under-priced. In the on-site bar, Tillingham's 2020 Chardonnay sells for £56 a bottle. But it was picked as one of her wines of the year by one of the world's most highly regarded critics, Jancis Robinson.

I didn't love everything I tried but the best have a freshness and purity that exemplifies the natural wine style. Tasting the pinot noirs out of tank was particularly fun as the previously taciturn Walgate came to life. He said he's getting 'more and more confident with reds. They used to be 100 per cent crushed and destemmed. Now we're moving away from that.' He's using whole bunches of grapes, fermenting with stems as well as skins. The 2021 vintage made in this way was notably spicy and herbal, in a good way. 'Received wisdom can fuck off!' he says. Rock 'n' roll!

If Walgate is the enfant terrible of natural English wine-making then Will Davenport is the eminence gris. In an industry that only really got going in the nineties, Davenport has been there from the beginning – though you'd never guess from looking at him. He's probably in his late fifties but could easily pass as twenty years younger. His winery and some of his vines are in East Sussex but most of his vineyards are a few miles over the border at Horsmonden in Kent, which he used as the name of the flagship white wine he has been making since 1991. It's a very English blend of various unfashionable German varieties which is consistently one of the most delicious and best-value wines in the country.

Davenport began his career as a wine merchant in England in the eighties, later studying for a post-graduate diploma in oenology at Roseworthy College in Australia ('it was either that or UC Davis because I didn't have the languages'). When he came back to England, his first job was at a vineyard in

Hampshire. He recalls that the grapes 'tasted amazing, much better than the flabby chardonnay in Australia. Australians would kill for grapes like that.' What he couldn't understand was why the wines tasted so bad. He tells a story about his wine merchant days when a colleague won a case of English wine and just gave it away. 'Nobody wanted it,' he says.

In 1991, he decided to plant his own vines at his parents' apple farm in Horsmonden, Kent with a view to selling grapes on to winemakers, but very quickly changed his mind and decided to make his own wine. He had a stroke of luck in 1993 when the Hampshire vineyard at which he'd been working went bust and he managed to acquire the winemaking equipment for a song. Straight away he decided he wanted to do things a bit differently. Initially he planted the standard Germanic varieties like ortega, huxelrebe and bacchus but rather than make the usual approximation of sweet German wine, he made a crisp dry white using wild yeasts, no de-acidification, and only a little added sugar to bring the alcohol levels up. The Horsmonden white has a depth of flavour that is still startling and must have been a revelation 30 years ago. Since then he's added superb sparkling wines made with French varieties, some great pinot noirs and more outlandish things like a pet nat. There's only one way to make wines like this – with fully ripe grapes. With natural winemaking, there's nowhere to hide.

I first tried Davenport's wines at the Natural Wine Fair in London in 2012. The show was organised by Doug Wregg of Les Caves de Pyrene (the distributor that Davenport shares with Tillingham) who is an evangelist for natural wines. Despite this, Davenport is ambivalent about being included in the scene. 'I don't like the term "natural wine". If you focus on the vineyard then you don't need to do much,' he says. 'People think natural wine is funky. We want our wine to taste how you expect wine to taste. It has to sit on a supermarket shelf.' Upmarket supermarket chain Waitrose is one of his biggest customers.

Davenport has now outgrown his Sussex site and is planning to build a new winery over the border in Kent so he can be close to most of his vineyards. From being something of a lone eccentric, Davenport has become extraordinarily influential on the English winemaking scene. He's the Yoda of natural winemaking (if Yoda was played by an early middle-aged Dennis Quaid). Talk to producers who are trying to do something a bit different in English wine, and his name comes up again and again. Davenport joked that he should have a get-together with all his ex-employees but he worries he wouldn't be able to remember them all. Winemakers such as Peter Morgan (ex-Nyetimber), Ben Walgate (Tillingham), Ben Witchell (Flint in Norfolk) and Adrian Pike (from Westwell in Kent) have all passed through his doors.

Westwell's Pike was inspired by a bottle of Davenport Horsmonden at a Soho restaurant – it was the first English wine that he really loved. He was working in the music business at the time but was losing his love for the industry and looking for another project. He had already picked up a passion for wine, routinely stopping in Burgundy to stock up on the way back from the annual music business conference in Cannes, and decided to work with Davenport to learn how winemaking was done.

Westwell was a producer near Ashford in Kent making sparkling wine but the owner was looking to sell. The company was bought by Rod Taylor, an electronics engineer, who took on Pike as manager and winemaker. Taylor died in 2018 but the family are still involved and Pike works most closely with Rod

CHAPTER 11

Organic growth

'People think organics is hippies walking around the vineyard not really knowing what they're doing but in fact it's really scientific. We're always looking for the latest cutting-edge ideas'

Will Davenport

Will Davenport's vineyard on the border of Kent and East Sussex doesn't look like the carefully manicured rows you see in California or the more expensive bits of Tuscany. The area around the vines is full of grass and straggly weeds. With the vines dormant when I visited in March, and the rain coming in heavily, it's hard to imagine that these unhappy-looking bits of spindly wood will yield some of England's best wines.

Davenport is one of the very few growers in England to farm

organically. Unlike 'natural wine', 'organic wine'[49] is a term regulated by a body called the Soil Association, which determines what can and can't be used in the vineyard. Creating wines without herbicides – especially synthetic fungicides – is hard in England's damp climate. Weeds grow like, well, weeds, and fungi such as powdery mildew, downy mildew and botrytis can wipe out crops overnight.

Mr English Wine Stephen Skelton[50] doesn't think that it's possible to farm organically and make a living. Vines just won't produce enough grapes. He tells a story about interviewing one well-known organic producer in South Wales for his book and being thrown out as soon as Skelton brought up crop yields. Yet somehow Davenport has not only done it but built a sustainable business that has proved a beacon for like-minded people across the industry. How?

For the first ten years or so, he farmed semi-organically but still relied on synthetic fungicides. Then, in 2000, he decided to take the plunge and convert to fully organic viticulture. It was a very Zen-sounding German wine consultant who persuaded him to go the whole hog: '"If you get the soil in balance, you won't get botrytis,' he told me," Davenport recalls. It seems he was right. 'We still get a bit of botrytis but nowhere near the levels we used to have.'

Davenport explains how it works: 'The whole principle is that you look after soil. If you have really healthy soil, the vines will be healthier.' He has his soils analysed and then uses compost, manure and natural sprays made from things like

---

49 There's actually two terms: 'wine made from organically-grown grapes', which only regulates what goes on in the vineyard, and 'organic wine', which governs vineyard and winery practices

50 See 'The Bloody Awful Weather Years', chapter 2

nettles. In organics, the only permissible inorganic chemicals are sulphur and copper sulphate – both traditional remedies for fungus. In wet years like 2021, it's a struggle. Davenport and his team had to work double the usual number of hours nurturing the vines, removing infected leaves at the first sign of trouble and producing half the crop. Committing to organic viticulture is not something to be undertaken lightly. But it seems to have worked. A sample of Davenport's 2021 pinot noir tasted from tank was gorgeously ripe, with not a trace of green.

Davenport believes organics is misunderstood: 'People think it's a bunch of hippies walking around the vineyard not really knowing what they're doing but in fact it's really scientific. We are always looking for the latest cutting-edge ideas.' He mentions a scientist he's working with to trial a bacterial spray designed to provide protection against frost. When it comes to fertiliser, he can only use things like manure and compost rather than adjusting the soil with potassium and nitrogen that are used in conventional agriculture but which can be problematic – nitrogen can run off and poison water sources, while the overuse of synthetic fertiliser can increase yield but at the expense of quality and concentration.

Davenport is a lot more relaxed about weeds than winemakers at other vineyards are, hence his less than pristine vines. He views conventional viticulture as mollycoddling vines, whereas organic viticulture toughens them up. His mother also grows grapes, and he says that while her leaves are like lettuces, his are leathery and tough – not as pretty but far more able to deal with adversity. He tries to get away with only mowing around the vines, to keep down growth, two or three times a year.

Hugo Stewart of Domaine Hugo in Wiltshire used to make organic and biodynamic wine at Les Clos Perdus in Corbières, southern France. He came back to England in 2015 and planted one hectare of the family farm with Champagne varieties. The idea was to farm without synthetic chemicals just as he did in France: 'I was very nervous about doing it here but it's actually easier,' he says. Stewart is now in his mid-70s but the outdoor life clearly suits him and he has a shaggy mane that most men his age would kill for. He argues that organics only cause problems when you have rain and the night-time temperature is above 10°C. In 2021, a disastrous year for some due to the damp, cold summer, he managed to get a good crop.

'You have to be on top of it – once you have it [a fungal problem] it's too late,' he says. It's important to let the grapes get plenty of sun and air so he removes leaves. Sometimes he has to spray with copper to prevent downy mildew, something he is not happy about because, in excess, it is highly toxic. He uses sulphur to combat powdery mildew, which he doesn't mind doing since apparently the soils are deficient in sulphur so it's good for the vines. As for botrytis, he claims that organic farming puts less stress on the grapes, resulting in thicker skins that won't split and allow the fungus to take hold.

Another grower who thinks that organics in some ways makes things easier is Henry Laithwaite at Harrow & Hope near Marlow in Buckinghamshire. Laithwaite was prejudiced against organic wines previously, describing them as 'just an excuse to make bad wines', but has since changed his mind. He converted his winery to organic production in 2020 – 'a really stressful year' – and saw yields drop by 20 per cent, but managed to get a good yield in the tricky 2021 vintage and then a bumper crop in 2022. 'I love telling Stephen Skelton about my yields,' he said.

It takes a vineyard time to adjust to a new regime. Laithwaite's vineyard, which he used to keep weed-free by turning over the soil with a harrow, hence the name, is now full of weeds which he allows to grow alongside the vines as they don't get too high. Photos from 2019 show a startlingly different-looking vineyard with the soil all neatly turned over and not a weed in sight. 'I was like an old Frenchman – I wanted to see my terroir,' he jokes. 'I've learned more in the last three years than in the previous ten.' One of the advantages of organics in England, according to Laithwaite, is that because of the high rainfall and fertile soils, vines cultivated in the traditional, non-organic way can produce too much foliage and too few grapes. In contrast, organic cultivation results in vines that aren't too comfortable, which then produce fuller-flavoured grapes.[51] '[Using] organics helps temper vigour,' he adds. He also thinks that grapes grown organically simply taste better: 'Organics produce flavour ripeness sooner.'

At Domaine Hugo, Hugo Stewart also thinks that growing grapes alongside other types of plants can help. One field of newly planted vines at his estate is covered with different species. 'Rather than competing, they are helping each other. Monoculture doesn't do any good.' Especially important are legumes like yarrow and chicory which add minerals to the soil. He lets the weeds grow up to three feet tall and only mows in July and August because of the danger of the cover crops (those planted merely to cover the soil rather than for harvesting) harbouring fungi. The cut plants form mulch around the vines

---

51 As Mark Gaskain, whose family has been farming fruit in Kent for generations and is now working with Taittinger at Domaine Evremond near Faversham, puts it: 'Happy plants grow leaves and wood. If you withhold water and nutrients just enough, they worry about the future and feel the need to have children to produce grapes.'

which in turn fertilises the soil and inhibits plants growing directly under the vines. It makes for a lively vineyard and although Stewart admits that the vines he planted two years ago might be thriving more without the competition, in the long run he thinks his system will prove better for the soil.

Working without synthetics, as in organic viticulture, might be a lot more work but for smaller vineyards, it can be cheaper. At Tim Phillips' one-acre walled garden which he calls Le Clos du Paradis[52], he does everything by hand and reckons that his outgoings come in at little more than £100 per year, mainly on petrol, copper sulphate and wood for posts to hold up the wire to which the vines are trained. Conversely, he does have to put in a lot of 'blood, sweat and tears', as he put it. It's been quite a learning experience. He has had three bad vintages in the last decade – 2014 and 2015, which he puts down to 'cock-ups', and 2021, which was a terrible vintage almost everywhere.

There are those, however, who are sceptical about aspects of organic viticulture. Another Davenport protégé, Adrian Pike at Westwell, says he is moving 'towards' organics but he remains concerned about botrytis, a fungus that thrives on sugar, affecting his thin-skinned ortega grapes, so he uses synthetic fungicides. Such treatments for botrytis, along with powdery and downy mildew, are relied upon by the English wine industry. Without regular spraying, you could lose your whole crop, especially during cool, damp summers like 2021. So important are chemicals like mancozeb and metalaxyl that Henry Laithwaite's father, Tony, founder of the biggest mail-order

---

52 Clos means walled vineyard in French, and the word paradise comes from a Greek word for walled garden

wine retailer in Britain, puts part of the success of English viticulture in the last 30 years down to improvements in fungicides (not quite as compelling a story as global warming).

The organic treatment for combatting fungus is sulphur and copper sulphate, a combination known as Bordeaux mixture. The downside of this is that large amounts of copper poison the soil, so it has to be used sparingly. (Davenport argues that he doesn't need to use too much copper because of the health of his soils.) Rob Saunders, an agronomist with agricultural supplier Hutchinsons, says: 'Excess copper that accumulates in the soil will have an adverse impact on soil organisms.' He is somewhat scathing on what is and isn't allowed in organic viticulture: 'Any technology available to grandfather is OK,' he says, sarcastically.

Ben Witchell at Flint in Norfolk told me that his life would be a lot easier if he was organic since, every day, visitors to the vineyard ask him why he isn't. His answer? 'Copper sulphate. It's very harmful.' As a compromise, he employs an electrostatic sprayer, which disinfects the vine but uses 20–30 per cent less chemical fungicide. 'I prefer synthetic [treatment] as it's more effective and you can use less.'

For Duncan McNeill, who looks after a number of vineyards in Essex, it's a question of knowing when to take action so you pre-empt problems and don't have to spray so much. Fungicides are the only synthetic treatments that McNeill uses. 'I don't use synthetic fertiliser as it kills bacteria, and makes the vines reliant on fertiliser. There are enough nutrients in the soil.' Rob Saunders cautions, though, that each crop harvested will remove relatively large amounts of potassium and, to a lesser extent, nitrogen from the vine. The nitrogen could come from synthetic fertilisers or from organic matter like chicken or livestock manure. He told me that vines can thrive for five to ten years without fertiliser but 'then they get hungry and performance will drop off'.

At Gusbourne in Kent and Rathfinny in East Sussex they have sheep in the vineyards which not only fertilise the soil but keep the weeds down (though the sheep are dispersed once the grapes begin to grow, in case they decide to eat them). They also look splendid. Kresse Wesling, who has just planted a vineyard near Faversham in Kent, uses sheep too. Along with her husband Elvis (yes really) she has come up with a system harnessing electric fences whereby the sheep are penned into a specific row so they eat *all* the weeds, not just the most tender, delicious ones. The fledgling vines are carefully protected with plastic tubing. She told me that they tried using goats but they were too naughty and kept on escaping. Hugo Stewart is working on a similar system and uses special Shropshire sheep which 'keep their heads down' (apparently some sheep breeds have goat genes and jump all over the place).

While not all organic farmers are concerned about weeds, they can be a problem, especially with young vines. Weeds compete with the vine for nutrients in the soil and water, reducing yields, and they can harbour fungi. If you don't have sheep to manage the weeds, the choice is between using herbicides or keeping the weeds down mechanically. Even here, there's some debate about the best way to do this and there are trade-offs at every stage. The problem is that all the alternatives to herbicides – ploughing or using machines called under-vine cultivators (basically fancy strimmers) – involve tractors, which have a financial and environmental cost. At a local level, they compact and turn over the earth, which according to Adrian Pike, 'is bad for the soil, and worse than herbicides'. Furthermore, tractors require diesel which is expensive and releases more carbon dioxide into the atmosphere, going against one of the goals of sustainability. Instead, Pike uses herbicides once a year. Duncan McNeill in Essex admits that he was 'burning more diesel' in not using herbicides. Rob Saunders comments: 'I have suspicions that the total environmental impact of cultivation is worse than [from] glyphosate herbicides. [By ploughing] you are disturbing the population of communities of microorganisms and damaging soil structure.'

Worries about the impact of industrial farming and synthetic chemicals on the land are nothing new. They were current when the seeds of the modern English wine industry were planted in the seventies. One of the pioneers of English wine, Gillian Pearkes, warned about it in her 1981 book *Vinegrowing in Britain*: 'The Champagne winefield is the only major wine-producing area, together with that of Beaujolais, to have adopted total herbicide weed control, and to have dropped cultivation . . . The chemical Round-up has largely replaced Simazine and Gramoxone, with many growers in Champagne being especially effective in controlling the more pernicious perennial weeds. This practice also contributes towards the unhealthy, dull-grey soil colour... The soil is solid and impacted due to man and his vehicles constantly passing up and down the rows.'

Champagne is notorious for its bare vineyards denuded of all life and 'fertilised'[53] with the rubbish from Paris which includes bits of plastic, glass and cigarette ends. American wine

---

53 This was a practice that started in the early 20th century when the rubbish from Paris was used to fertilise Champagne. This was fine when all people threw away were potato peelings and chicken bones, but it continued as people's rubbish began to contain more plastic. The practice was stopped in the 1990s but you can still see little flecks of plastic in the famous Champagne terroir

writer Alder Yarrow wrote: 'Many of Champagne's most storied vineyards are seemingly lifeless other than the vines that emerge from the soil. And to add insult to their chemically denuded injury, they are quite literally covered in trash.' It's not just in Champagne. Wine consultant Claude Bourguignon once said there is less life in some Burgundy vineyards than in the Sahara desert.

This is usually blamed on herbicides. McNeill told me: 'Herbicides kill bugs and sterilise soil. That doesn't sit well with me.' Most wine writers too are very anti glyphosate, the best-known brand of which is Monsanto's Round-up. One commentator, Libby Brodie, posted pictures of herself on Twitter next to a vineyard that looked like it had been scorched from heavy herbicide use. The critic Jamie Goode opined about 'ghoulish zombie vines rising out of the dead, stained earth,' while Aaron Ayscough went further and described it as 'like vogue-ing in an oil spill'. Meanwhile over on Jancis Robinson's website, Tamlyn Currin writes: '[Talking about] conventional [farming] is, plain and simple, whitewashing a crime against the planet in my view. We should, instead, be using the much more ugly, but much more realistic term "synthetic-chemical viticulture".' Blimey!

When it gets as heated as that it's refreshing to hear from people like Keith Brennan, a farmer in Ireland who stated: 'I will take a good conventional farmer over a bad organic one. I've met many good organic farmers and many good conventional ones. They have much in common.' Rob Saunders, who, as you'd expect, comes down firmly on the conventional side, thinks the problems of herbicide use are either exaggerated or misunderstood. He explains that only very small amounts of glyphosate make it into the soil, where they are easily broken down. What does damage the soil, however, is the removal of plants themselves: 'If you remove the green material from the surface of the soil you will impact soil biology, whatever you use.' The lifeless vineyards that you see in Champagne are exacerbated by constant compaction from tractors, while the lack of plants means that the fertile topsoil is washed away, leaving the vines to rely on synthetic fertilisers.

Saunders' advice is to plant slow-growing crops which need infrequent mowing, and to then use minimal herbicide directly under the vines. Legumes such as vetches are especially good as they fix nitrogen into the soil. He cites the example of Jon Pollard, vineyard manager at Gusbourne in Kent, who uses a single application of herbicide each year, with some cultivation. Chris Foss is another who thinks glyphosate has been unfairly maligned. After retiring from Plumpton College, he now heads up the Wine GB Sustainability Scheme which was launched in 2019. He described glyphosate herbicides as 'not so bad'. 'One person died of cancer because of glyphosates, Monsanto paid out £1m and made a lot of noise. Look at the figures, it's not as bad as all that.'

Foss also thinks we're quite lucky in England in not having the severe insect problems that they have in warmer countries. The spotted wing drosophila, which has come over from Japan, is causing headaches and Charlie Holland from Gusbourne says this kind of fruit fly, with its 'drill bit' that can burrow deeply into fruit, has been a problem since 2007. Gusbourne uses targeted insecticide when it discovers an outbreak. For Davenport, the answer is to build a biodiverse insect-and-bird population to control such outbreaks. Larger animals like deer, rabbits, pheasants or badgers can also be a pest. 'Badgers eat grapes and then shit everywhere,' says Adrian Pike at Westwell.

163

Fiona Wright from Langham sympathises: 'The worst thing are the badgers, standing on their hind legs eating grapes.'

Wine GB's Sustainable Wines of Great Britain scheme requires applicants to adhere to three key practices in their winemaking: manage vineyards sustainably with minimum pesticide and fertiliser input; reduce water and non-renewable energy consumption and minimise carbon footprint; and protect vineyard soils, conserve the environment and promote biodiversity. Anyone can sign up – viticulturalists just have to record what they do and then move in the right direction by cutting down on pesticide use. Although the organisation is audited by an independent company, Ricardo, nobody is going to come round to vineyards with a clipboard. As Foss puts it, the scheme 'relies on trust and honesty'.

Most of the big names in English wine such as Chapel Down and Ridgeview are onboard, but the scheme has come in for criticism. Neither Davenport nor Flint has signed up. Mark Driver of Rathfinny in Sussex says you could 'drive a bus through the regulations'. Instead he's pursuing B-Corp certification, to show the company's credentials as a leader in social and environmental performance (something we'll look at in closer detail in chapter 18). Meanwhile Tom Barnes at Biddenden in Kent thinks the scheme is really aimed at producers who sell wine in supermarkets so they can put a label on their bottles. 'Why do I need to pay to say I'm sustainable?' he asks. 'Everyone can see what we're doing here.'

Foss admits that there is a marketing angle to the scheme: 'We are selling an experience, a culture which has to include sustainability if we're going to do it properly.' Ben Walgate told me that his views on the Wine GB sustainability scheme should not be published. 'It's not far from a greenwash,' he said. 'The techniques held up to be entry level are woefully below what people should be doing.' Adrian Pike laughed when I mentioned Foss and the scheme, referring to Foss as 'Chemical Chris'. 'He wouldn't go near organics and we never heard of sustainability when he was at Plumpton,' he added. Yet Pike has signed up to the scheme, reasoning that it's better to be influencing things from within than to be outside on the fringes criticising.

One of the main problems with the word 'sustainability' is its very vagueness. It's about using fewer synthetic chemicals and being less damaging to the environment but as we have seen, there can be a contradiction between trying to lower carbon emissions and not using herbicide. Organics can be much more energy-intensive while, for some, the word 'sustainability' tips into HR matters like equality, diversity and staff welfare. In trying to do so much, sustainability risks achieving nothing while overburdening a young industry with regulations. The other issue is a richly deserved cynicism about large companies using the word 'sustainability' as a marketing tool. And finally, does it mean anything to customers? Will being part of the Wine GB sustainability scheme help shift bottles?

But whichever way you do things, there does seem to be an agreement across the industry that soil health is important in ways that we are only just beginning to understand. In the past it was thought that soil was essentially just a bed of nutrients 'but it's much more interesting and complicated than that', says Saunders. 'The soil is made up of a massively complex food web of bacteria, fungi and protozoa.' As New Zealand winemaker James Millton[54] puts it: 'We're not standing on dirt, but the rooftop of another kingdom.'

---

54 Quoted in Jamie Goode's *Regenerative Viticulture*

This growing awareness of the importance of soil health has come together in a loose movement called 'regenerative viticulture'. This was a term I'd never heard until I visited Ben Walgate at Tillingham. I was hoping to have lots of fun with him debating the efficacy of various biodynamic practices like burying cow horns in certain parts of the vineyard but he batted away my questions. 'It's not that complicated – healthy soil life means a better immune system in the plants and more profit.' He continued: 'To farm the right way, sustainability barely scratches the surface. We need to wake up and smell the coffee.' Again and again on my travels around England I came across those two words – regenerative viticulture or 'RV'.

Justin Howard-Sneyd is a former buyer for Waitrose supermarket who now makes wines in the Languedoc as well as consulting with various English vineyards as a trustee of the Regenerative Viticulture Foundation. He describes RV as 'a mindset about soil health. Soil is a bank account that we are trying to build up, whereas [until now] we have just been making withdrawals.' Bad practices include ploughing and turning the soil this disrupts habitats, plant life on the surface is killed and the fertile topsoil is exposed to wind and rain. This is why estuaries and rivers are silting up more than in the past, he says. The more soils are denuded, the more farmers will rely on chemical fertilisers which depend on fossil fuels, whose use is not only unsustainable but increasingly costly. 'We're going to end up with empty bank accounts,' says Howard-Sneyd.

But soil health is a complicated business. Some producers associated with RV do till their land and have found that it increases the health of the soil. Rather than use herbicide, Christina Rasmussen uses a small electric tiller at her small vineyard in the Cotswolds, which she only planted in 2021. She hopes that when the vines are more established she'll be able to move away from tilling but she's not sure. One grape variety in particular, mondeuse, is struggling with the competition from other plants. She cites some vineyards in Burgundy with heavy clay soils where growers have found it essential to turn over the soils to aerate them. For her it's about getting to know what works best in her vineyard. 'There are so many approaches that are too prescriptive,' she says. 'A consultant will come in and do what he did in another region. But I ask, "What does this vineyard need from me?"'

The RV movement is split on the issue of herbicides too. Simon Porter has experience in regenerative arable farming as well as growing grapes in Hampshire. 'Glyphosate has become so political,' he says. 'There's a vendetta to get it stopped.' He argues that 'Now that we have glyphosate, never before has mankind had the ability to safely produce so much food with the least amount of harm.' Porter tells a story about digging up some soil to show a group of students from Plumpton College how full of life it was. One student accused him of not digging up soil that had been treated with Round-up which, he claimed, 'kills everything'. So Porter extracted some soil where the weedkiller had been applied, and it was exactly the same.

There's evidence that such herbicides only kill the plant they touch and then break down into the soil – though many of these studies, Howard-Sneyd recognises, are funded by chemical companies. But on balance he thinks, 'It's better not to plough and to use a little weed killer. But there are people who disagree. The message is a nuanced one.' There are stories about herbicides causing hermaphroditism in frogs. Wesling says 'You can't talk to me about terroir and then use glyphosates.'

RV involves treating the soil, the vineyard and the

surrounding area as an entire ecosystem, not unlike bio-dynamics. But unlike that specialist practice, it's based on what works for each site rather than being proscriptive. Howard-Sneyd continues: 'RV is not about the things you can't do, like organics. It's a path to understanding how things work. It's open to scrutiny and, unlike biodynamics, is scientific. I don't like the lack of openness and scrutiny [in biodynamics].' Bio-dynamics is a system of agriculture that was invented by an Austrian mystic, some would say crank, Rudolf Steiner, despite a lack of practical farming experience. Tim Phillips doesn't farm according to strict biodynamic principles and describes Steiner as 'a thinker rather than a doer', but he does find some of his recommendations useful, citing dandelion, or preparation 506, as being particularly so.

The problem with biodynamic viticulture for non-converts is that in among the practical stuff there's a lot of new age woo woo. Demeter, the biodynamic certification's governing body, advises stirring herbicidal preparations in a special way in order to 'release the dynamic forces they contain and transform them into a rhythmic activity which can then stimulate a cor-responding activity in the growth and development of the plants in the soil'. Caroline Gilby MW[55] writes of a biodynamic preparation for rodent infestations where, after the capture of field mice, practitioners should 'skin them and burn these skins when Venus is in Scorpio, then sprinkle the ashes like pepper across your fields. And then after all this, these substances are used in miniscule quantities – the contents of one skull treats 300 hectares of land.' As you might have guessed, she is something of a sceptic: 'I remain to be convinced that most of

---

55 thewinesociety.com

it is anything more than mysticism, moonshine and marketing hype overlaid on top of some sound and pragmatic grape-growing practices.'

Even biodynamic practitioners are sceptical about its more outlandish elements. Like Ben Walgate, Howard-Sneyd says that 'most people do the practice because they have to. It works, but not for the reason they think.' Dan Ham at Offbeat compares biodynamics to ancient Chinese medicine, in that 'it's holistic'. But there's also a lot in it that Hugo Stewart describes as 'peasant knowledge'. For example, he always prunes the vines when there's a full moon so that the sap is rising (sap flow activity in plants is proven to increase with exposure to light) thereby ensuring the cuts are full of sap and so reducing the danger of the vines dying where they have been cut. He admits there's also a lot of 'muck and mystery... like homoeopathy'. 'Lots of people poo poo homoeopathy. Sometimes it works. But you can't prove it.' Wesling describes biodynamics as being like yoga, in that it contains a spiritual element, whereas regenerative viticulture is more like pilates, as it's scientific.

The RV dream, Howard-Sneyd concludes, is 'the truly healthy plant that protects itself'. And this might mean moving away from vitis vinifera (the classic Eurasian family of vines). 'We import delicate vines that are not suited to the climate,' he says. 'We have bred them to create flavour and grown them to be addicted to fertiliser and anti-fungal treatments.' Tim Phillips describes himself as a 'classicist' for his love of chardonnay, sauvignon blanc and riesling, but he acknowledges the insanity of the amount of work needed to grow such varieties in England. He reckons it takes about 1,000 hours of labour per hectare of vines compared with about 40 minutes of labour per tree for apples (not including the harvesting).

Howard-Sneyd thinks the future is 'more robust plants' which don't need all those treatments.

There's one vineyard at Chew Magna near Bristol that has taken this to extremes. Look at photos of Limeburn Hill Vineyard online and it looks as if it has never been weeded, the vines competing for space with all kinds of other plants. It's enough to give an old-school vineyard manager a heart attack. But rather than hinder the grapes, owner Robin Snowdon[56] thinks that the other plants help by introducing certain chemicals into the soil at the right time in the growing season. So dandelions, which are high in potassium, grow early in the season and vines need potassium for developing flowers and fruit. Yarrow, which is high in sulphur, a natural anti-fungicide, thrives in the summer when grapes need protection from mildew.

Snowdon is able to take this approach by using particular grape varieties called PiWis (from the German *Pilzwiderstandsfähig*, meaning fungal resistant). He's not the only one. Barry Lewis, who makes wine at Amber Valley in Derbyshire, writes on his website[57] of managing to drop the use of fungicides entirely by using varieties that have been specifically designed to thrive in colder climates. Kresse and Elvis Wesling too have planted only PiWis at their experimental vineyard near Faversham in Kent. Their main business is turning old fire hoses into luxury handbags so Kresse feels she has some experience of alchemy. Could such varieties be the future of English wine?

---

56 internationalwinechallenge.com
57 ambervalleyvineyards.co.uk

CHAPTER 12

Grape expectations

'Personally, I've never enjoyed wines produced from these varieties. They're challenging to drink or downright unpleasant'

Sergio Verrillo

When you first start learning about wine these days, it's grape varieties you are told to focus on. You learn how cabernet sauvignon smells of black-currants, sauvignon blanc is redolent of green peppers and gooseberries, and gewurztraminer smells of lychees. Although come to think of it, I don't think I've ever actually tasted a lychee. So ingrained is the idea that particular varieties have certain flavours by which you should categorise wine, that it's easy to forget that this is a very modern way of looking at things.

In the not so distant past, where a wine came from was all important. You might love red Burgundy or Bordeaux but you'd have to be a specialist to know that the former was made with pinot noir and the latter cabernet sauvignon and merlot. In fact, growers may not have even known which varieties were planted in their vineyards – it would have just been a 'field blend' of the local grapes. There are still vineyards like this in Portugal and other countries.

Despite the tyranny of varietal spotting, geography is still a helpful way of thinking about wine. To take one example, pinot gris and pinot grigio are the same grape, but a pinot gris from Alsace will taste much more like a gewurztraminer from the same region than it will a pinot grigio from Northern Italy. Furthermore, grape varieties are prone to the vagaries of fashion. In the eighties and nineties the traditional varieties of the Languedoc, like cinsault and carignan, were considered only suitable for plonk. There were EU grants to pull up old vines of unfashionable varieties and replace them with syrah and cabernet sauvignon. It was a similar story in South Africa, Lebanon and Chile. Nowadays, however, old-vine carignan and cinsault from these countries are highly prized. Some of the best Cape cinsaults sell for more than £40 a bottle.

All this preamble is relevant to English wine, I promise. When thinking about the grapes that grow in England, it's important to undo years of prejudice about what are superior and inferior varieties. Any variety can be maligned and then undergo a reappraisal. What is often more important than the variety is how and where it's grown, and the age of the vine. That's without going into how the wine is made. In this book there will be varieties you have probably never heard of. There will even be vines that, because they contain American vine DNA, are not allowed to go into quality wine in Europe. But we shouldn't dismiss them until we've tried them. This is a mistake I've made time and time again.

There are, broadly speaking, three types of vines growing in England today: French varieties, mainly those grown in Champagne and Burgundy; Germanic varieties, which are usually crosses created by scientists to work better in colder climates; and hybrids made by crossing vitis vinifera[58] with other types of vines. There are also some tiny amounts of unexpected things like the Iberian variety albariño grown in Kent.

Until recently, the backbone of the English wine industry were varieties like müller-thurgau, huxelrebe, reichensteiner and schönberger. As we have seen in the 'Bloody Awful Weather' chapter, the first crop of trained English winemakers looked to Germany for inspiration. Consultants like Karl-Heinz Johner and Stephen Skelton were extremely influential. John Atkinson from Danbury Ridge refers to Skelton as the 'emissary of Geisenheim', the German wine institute. There, growers had crossed famous German grapes like riesling and sylvaner to create new varieties that ripened more reliably in Germany's marginal climate – most notably müller-thurgau. These new varieties were responsible for the cheap sweet wines like Blue Nun and Black Tower that were massively popular in England in the sixties and seventies. And yet most of Germany's vine-growing regions have warmer, drier growing seasons than south-east England. The problem these varieties had in Germany was that they ripened too quickly, creating plenty of sugar but very little flavour. They were also cropped with massive yields – and lots of grapes per vine is a recipe for low-quality fruit.

In England's more marginal climate, however, these varieties tend to ripen more slowly, which, at least in theory, means that if you can get them fully ripe, you will harness more interesting

---

58 The classic Eurasian vines that make most quality wine worldwide

flavours. Sadly, in the past, worries about disease and colder weather meant growers would pick early and make up for a lack of ripeness by adding sweet grape juice known as sussreserve. It's a shame, since if you get properly ripe grapes and ferment to dryness, these varieties have the potential for high-quality wine. I've had some lovely blends made from these German varieties. The problem is that nobody wants words like müller-thurgau or reichensteiner on the bottle. They don't appeal to customers brought up on pinot grigio and sauvignon blanc. French and Italian varieties sell. Germans don't.

There are two exceptions to this rule, perhaps because they don't sound German: bacchus and ortega. First planted at New Hall in Essex, bacchus has become the backbone of the English (still) wine industry, producing grassy aromatic wines that are redolent of sauvignon blanc. It's a tricky variety though, because if it gets too ripe it can taste over-pungent and you need to be careful which yeast you use to ferment it. Most producers blend ripe and overripe grapes to create a balanced style. It's usually quite a basic wine but in the right hands and especially if it spends some time in old oak it can be very good. Ortega, mean-while, seems to work brilliantly in parts of Kent. It's a rare example in England of a variety that has found its perfect home. The Barnes family at Biddenden produce a tasty off-dry version while at nearby Westwell they make something like an English vinho verde.

There are also various German red grapes like dornfelder which hold out the promise of making good affordable red wine in England. Tom Barnes from Biddenden refers to dornfelder as the 'bank manager's grape' – it produces such opulent bunches that it's what a winemaker shows off when they're asking for a loan. That's the thing about the less-fashionable varieties – they often produce higher yields than sexy French grapes like chardonnay or pinot noir. This means cheaper wines. You're never going to get a bottle of good 100 per cent English pinot noir for much less than £25, but by blending in obscure, higher-yielding varieties, it's possible to make something like New Hall's Barons Lane red which costs around £14. It's a blend of various crosses combined with pinot noir precoce[59] and zweigelt, an Austrian grape that I'd love to see more of.

For reasons that are not entirely understood, old vines tend to make better wine. Many of those unloved German varieties planted in the seventies and eighties are now in the prime of their lives and should be producing excellent fruit. One Master of Wine, Tim Wildman, has started a project to find and rescue old, neglected English vineyards and make wine out of their fruit, rather like small bands of growers in France, South Africa and beyond began to treasure their old varieties as they were being pulled out. He produces the Lost in a Field 'Frolic' pet nat from what he calls 'English heritage vines'. His worry is that as growers rush to plant more fashionable varieties like pinot noir and chardonnay, something precious will be lost. 'Some of these varieties literally don't exist anymore,' says Wildman. 'They're extinct. They're the dodos, the dinosaurs of wine. The nurseries don't have them anymore. So, the madeleine angevine that's planted in England, for example, is the last.'[60] But sommeliers such as Donald Edwards at La Trompette in west London think that the pendulum is swinging away from French varieties to some of these neglected gems. The moral of the

---

59 An early-ripening clone of pinot noir
60 Interview with *Club Oenologique* magazine

story is that unfashionable varieties can become fashionable again very quickly.

The great white whale, however, of Anglo-German wine is to properly ripen riesling, the ultimate cool-climate grape. The problem is that it ripens very late so needs a long growing season with plenty of sunlight and ideally a dry climate, something you get in a continental climate like Alsace and Germany but sadly not in England. As far as I can tell the only person to have ripened it successfully in this country is Tim Phillips from Charlie Herring wines in Hampshire. His vineyard is in an old Victorian walled garden that was specifically designed to ripen exotic fruit and vegetables. It was hard to tell on the freezing January morning when I visited, but in the summer and autumn the garden is apparently noticeably warmer than the surrounding area. He thinks within the walls he gets 1200 growing degree days[61] [a measure of heat accumulation used by horticulturalists to predict when a crop will reach maturity] compared to 850 outside. But even with the help of the walls, it hasn't been easy managing the grape's famously high acidity. In some vintages, Phillips has to blend in some chardonnay to soften it or make a sparkling wine.

Nonetheless, he seems to have cracked it. 'Riesling really is royalty on this site,' he says, 'quietly producing incredible fruit with minimum fuss.' In 2020 he managed to produce a highly-regarded 100 per cent still riesling described by wine writer Lisse Garnett[62] as 'spectacular'. Phillips said demand outstripped supply by a factor of six, and sadly the wines were all sold by the time I visited. The 2022 vintage is similarly promising

---

61 See glossary
62 *Wine Anorak* blog

though. The riesling vines looked incredibly delicate next to Phillips' rampant sauvignon blanc, but apparently it grows 'like a dream'. So much so that if he were to do it all again, he says, he'd replant solely with riesling.

Phillips is in the minority, though. We saw in the Money Men chapter how Mark Driver at Rathfinny eventually pulled his riesling up as he couldn't get it ripe enough, while also in Kent, Tom Barnes at Biddenden remains optimistic of one year ripening his riesling fully – but he hasn't done it yet.

While English growers originally looked to Germany for inspiration, long ago there were voices pointing out that France was a better model. Gillian Pearkes wrote in the eighties:[63] 'Not only because Champagne is the nearest wine region to Britain, is it the area from which we have most to learn. In Champagne they grow the same vines as some of us attempt to, and they suffer just as much with the weather.' Despite the presence of some chardonnay plantings in the eighties, it took the success of Nyetimber and Ridgeview to firmly turn the industry away from Germany and towards France.

The main French varieties grown in England today are chardonnay, pinot gris and pinot blanc, plus some others like auxerrois, gewurztraminer and sauvignon blanc,[64] and the Swiss chasselas. Most of these are harnessed for sparkling wine, along with the red varieties pinot noir, pinot noir precoce (an early ripening clone of pinot noir also known as fruhburgunder) and pinot meunier. But there's now a move towards making quality still wines from these varieties too, with varying degrees of success. In the last five years, standards of still chardonnay have

---

63 *Vinegrowing in Britain*
64 At the WineGB tasting in 2021 I tasted three sauvignons from England, all of which were introduced as the only sauvignons in the room

rocketed in line with the ripeness levels. I've also had some extremely tasty pinot gris, chasselas and pinot blanc, which seem to have a particularly bright future. We're now starting to see some increasingly good reds made from pinot noir and pinot meunier (though both can also be made into white wines). In warm years, Biddenden makes a light, fruity gamay, the Beaujolais grape, which clearly has a bright future in England's warming climate.

While most English winemakers are looking to Champagne, Burgundy or Alsace for inspiration, Christina Rasmussen has planted vines from the Jura, the region next to Burgundy, at a small vineyard in the Cotswolds which boasts savagnin (a clone of gewurztraminer), pineau d'aunis, mondeuse and trousseau. It will be interesting to see how she gets on. She is yet to have her first harvest but she's confident about the savagnin, which seems to be thriving, while the mondeuse isn't doing quite so well (she describes them as like the hare and the tortoise). She's also planted some more conventional chardonnay and pinot noir.

What these last two grapes have, beyond popular flavours, is name recognition; they can be ordered confidently by people who know very little about wine. From a very low bar in the nineties, these grapes now dominate the English wine industry. Their disadvantage is that while most years they can be ripened sufficiently to make sparkling wine, riper grapes are required for still wines. Only the very best still wines from these varieties can escape the dreaded English pinch, that hard acidity and tell-tale green note that lets you know that they were grown over here. Furthermore, yields are always going to be low, which means prices will be high.

In the vineyard, chardonnay seems to get every ailment

going, and Tim Phillips knows how difficult it is to grow. He describes it as 'like a beautiful girl who knows she's beautiful and causes so much trouble that sometimes you question whether it's all worth it'. It is when you get results such as Phillips' still sauvignon blanc (an 'unruly' variety, he says) blended with about 6 per cent chardonnay, the 2022 vintage of which, tasted from cask, was sensational. No wonder the wine has been compared favourably with those of Didier Dagueneau, a renowned producer in the Loire.

The grape variety question is further complicated by the existence of different versions of each variety, known as clones. You might think that chardonnay is just chardonnay but various clones have been isolated to do different things, such as resist disease or ripen early. Some taste bizarre: at Balfour in Kent, they have a clone of chardonnay with a distinct muscat-like, floral aroma. It's a similar story with pinot noir, where various German, Champagne and Burgundy clones exist. It can be hard to keep track of, but it basically boils down to the fact that some kinds of pinot noir might be great for making sparkling wine but are less good for still wines. Or some might be more disease-resistant but not taste as good.

Owen Elias from Balfour says that clones are vital to under-standing the transformation in English wine. 'People didn't know about clones until the 1980s when the French started identifying them,' he says, adding that growers would buy pinot noir but would have no idea if it was from Alsace, Champagne or Germany, or if it might even be pinot noir precoce, a mutation. 'Clones and site selection are crucial to what happened in English wine,' he says. In other words, finding the right grape for the right place. In the early days of the industry, clones weren't so well understood and there was talk of inferior versions of vines arriving in England. These might have been clones that produced poor-quality grapes or even vines that were diseased. Gillian Pearkes wrote at the time: 'There is no doubt that in many cases we are being sold very inferior stock, not only in terms of the quality and the chance of survival of the initial plant, but also in the strains and clones being planted in England.' Looking back, Peter Hall at Breaky Bottom agrees.

Nowadays you can buy vines by clone but originally such detail was only identified by growers, who would see a certain vine thriving, mark it with a bit of paint and then take cuttings from it when they needed to replant. Known as 'massal selection', this is a practice that Tim Phillips is returning to at his vineyard in Hampshire. The idea is that he propagates the vines that work best and are most resilient in his vineyard, and eventually he has clones that are much better suited to England. 'Come back in 300 years and see how I'm doing,' he jokes.

All the vines I have written about so far are vitis vinifera, the vine that probably originated in the Caucasus and has been cultivated all over Europe since Roman times. There are dozens of different species of vine, such as vitis labrusca from America and vitis pseudoreticulata from China, though most aren't suitable for making high-quality wine. Because of phylloxera, a vine-eating aphid that came from America and destroyed vineyards first in Europe and then all over the world in the late 19th and early 20th century, most vines are grafted onto American non-vitis vinifera rootstocks, which are resistant to the pest. This adds another layer of complication as some rootstocks work better in different conditions, and as with clones, English growers are only just beginning to work out where they work best. In the past, most rootstocks were selected for their vigour but some growers like Christina Rasmussen are using lower vigour ones, which in theory should produce fewer grapes, but of higher quality. Some isolated vineyards, like Tim Phillips', are able to work with the vines' own roots because they are phylloxera free.[65]

American species are also less susceptible to hazards like powdery mildew and can ripen in lower temperatures. The dream has long been to marry the taste of European vines with the hardiness of other varieties. So far, the results have been mixed but they are getting better all the time. One of these so-called hybrids is seyval blanc, which was created in France in 1921. Along with müller-thurgau, it used to be the backbone of the English wine industry and, like that variety, it has its critics. Owen Elias describes its flavours as like raw potato and cabbage. When he was at Chapel Down, he pulled up all the seyval blanc in Lamberhurst and Tenterden vineyards, and banned it from being replanted. Yet Peter Hall at Breaky Bottom produces some sublime sparkling wines from seyval blanc that can age for decades. As ever, don't knock it until you've tried it.

To make a reliable, deep-coloured red wine in England many growers turn to hybrid varieties such as regent or rondo. The latter, a much-maligned red grape with some native Asian heritage, is popular because it has red flesh, so it produces wines with lots of colour. I've never been much of a fan of its strange, rather confected flavours, which remind me of the fizzy drink Vimto. Nor has Sergio Verrillo from Blackbook in London: 'Personally I've never liked the wines produced from those varieties – they're either challenging to drink or downright unpleasant.' Owen Elias is not a fan either. 'Never liked rondo,'

---

65 Thanks to stringent quarantine, phylloxera, which ravaged Victoria and New South Wales, has so far been kept out of South Australia

he says, though he did accidentally once make a great one by mistakenly leaving it in cask for a year and then in bottle for three years. 'Make it like a Rioja,' is his advice, though he points out that 'it's not really economical to keep it for four years'. Yet other winemakers are making some nice wines with a rondo component. Flint in Norfolk produces a very nice sparkling pink that owes its pretty colour to the peculiar charms of rondo.

There's a certain class of hybrids, known as PiWis (short for the German Pilzwiderstandsfähig, meaning fungal resistant), which are particularly suited to the English climate and increasingly used in organic viticulture. One of the most popular ones, solaris, is planted at many of the more northern vineyards in England. Other PiWis are hybrids of classic varieties like pinot noir or cabernet sauvignon with non-vitis grapes, the theory being that they retain some of the character of the classic but with fungal resistance. Among them are sauvignac, which is made from riesling and sauvignon blanc; pinotin, based on pinot noir; and cabernet noir, based on, you guessed it, cabernet sauvignon.

Two of the growers we met in the last chapter have bet heavily on PiWis. Kresse Wesling has planted cabaret noir, souvignier gris and pinotin at her vineyard near Faversham in Kent. The idea is to grow them without any herbicides, insecticides or fungicides. The vines went into the ground in 2020 and the winemakers are expecting their first proper harvest in 2025. Meanwhile, Robin Snowdon of Limeburn vineyard near Bristol makes all kinds of weird and wonderful wines from his PiWis, including pet nats and orange wines (wines made from white grapes which are treated more like red grapes, via skin contact and therefore tannin).

Charlie Holland from Gusborne thinks that English winemakers are missing a trick not making more conventional wines from PiWis. 'No one's nailed it yet,' he says. He's looking forward to when someone 'uncouples beardy-weirdy non-sulphide from PiWi'. Holland, you might gather, is not a fan of natural wines. Sergio Verrillo isn't either, but he made a wine from cabernet noir in 2019 that he was pleased with, saying in an interview with the *Drinks Business*: 'The profile of cabernet noir is one that is reminiscent to cabernet sauvignon. I would say that the larger public body prefers cabernet sauvignon attributes and characters than those weird and wonderful hybrids [like rondo or regent].'

I've had some orange wines made from solaris which have been palatable if not outstanding but I haven't managed to try many of the newer PiWis. I'm trying to keep an open mind. It's important to remember that England is only at the beginning of its journey of discovery. France has had hundreds of years to work out which vines work best in which places. Australia and California have been making wine since the 19th century and they are only just figuring this out. No less an authority than Robert Louis Stevenson wrote, back in 1880, about how Californian vine growers were still at the 'experimental stage' but at some point would find 'their Clos Vougeot and Lafite. Those lodes and pockets of earth, more precious than the precious ores, that yield inimitable fragrance and soft fire . . . The smack of Californian earth shall linger on the palate of your grandson.'

Swap the word 'Californian' for 'English' and you have some idea where we are now. And in one small part of the country, in the damp clay of Essex, one producer thinks it may have found England's answer to Château Pétrus . . .

CHAPTER 13

Eastern promise

'Even though there are no maps to guide us, intuition tells me that somewhere out there in the "uneventful countryside" between Colchester, Epping and Burnham hides our Pétrus'

John Atkinson

If you want to see the future of English wine, go east. Essex is currently producing arguably the finest wines and inarguably the ripest grapes in England. But, perhaps because the very word 'Essex' sounds to most British ears like a punchline to a joke, or perhaps because it goes against the prevailing narrative in English wine, it's a story that is only just beginning to emerge.

During the early days of the English wine revolution, newspaper articles were full of comparisons between the soil in southern England and Champagne. There might even be maps showing the chalk seam that runs from France, through Kent,

Sussex and Hampshire and into Dorset. Thirty years later, this is still the story that sells. What nobody was talking about, until recently, was clay; thick, sticky London clay like you get in Essex. Such was the chalk mania of the past that Stephen Skelton, vineyard consultant, jokes he would have been chucked out of the Institute of Masters of Wine for saying he loved clay. Yet the English wine industry has a dirty little secret; much-lauded wineries like Camel Valley in Cornwall, Chapel Down in Kent and Lyme Bay in Devon have long been sourcing significant amounts of their fruit from Essex. Something like 40 per cent of Lyme Bay's grapes come from there, and when I joked with James Lambert, the head winemaker, that he should open an Essex winery, he gave me a look like he'd already thought of that.

Chelmsford's CM3 postcode has the highest concentration of vines in the country. The Dengie peninsula, the area between the Crouch River to the south, the Blackwater River to the north and the North Sea to the east, is literally and figuratively the hottest place to grow grapes in England at the moment. But until recently, you'd rarely see the name Essex on the bottle. It was a fruit-growing place that sold its crop: the big brand names like Chapel Down, Nyetimber or Hambledon were in Kent, Sussex, Hampshire or the south-west.

Lucy Winward works as commercial manager at New Hall vineyard near Chelmsford. She's one of those ridiculously energetic young people that English wine seems to be full of. After graduating from Plumpton College in 2013, she worked in Sussex at a time when, she said, wine from Essex did not have the best reputation. That might have had something to do with the scarcity of producers with large marketing budgets. Certainly there was nobody banging the drum for Essex like Hambledon did for Hampshire or Camel Valley for Cornwall. 'We have wonderful growers,' she says, 'but no one is very good at marketing. Crouch Valley should have been on the radar years ago.'

Essex's only big name was New Hall, which was founded in 1969, making it ancient by English standards. There aren't many vineyards in England with 50-year-old vines, but when I visited the estate, Winward pointed out a vast, gnarled marechal joffre[66] vine that looked like something from Lord of the Rings. There's also pinot noir that was planted in the 1970s. 'We don't know the clones,' said Winward. 'It was a back-of-a-lorry job.' There's also some zweigelt – a quality Austrian red grape that Winward thinks has great potential in England. One gets the impression that part of the fun of working at New Hall is discovering all the crazy stuff that's in the vineyard. But it was a white variety that sealed New Hall's place in the history books. The original owner Bill Greenwood was the first person to plant bacchus, a minority vine in Germany which has since become the workhorse of the English wine industry, after his son Piers brought the vine back with him having studied in Alsace and Germany.

Owen Elias, former winemaker at Chapel Down, was full of praise for the fruit that he sourced from New Hall. 'New Hall was an outlier, it's a really good site. They actually get pinot noir ripe there.' Even though the vines 'weren't well looked after and they cropped heavily, they still ripened two weeks before Kent,' he recalls. Elias made some nice, light pinot noirs from the fruit, but apparently the sales team didn't like them as they found pale reds difficult to sell.

---

66 Obscure French variety named after top First World War French general Marshal Joffre

As well as producing wine under its own labels, New Hall historically sold grapes and also produced contract wines for other local growers. The winery is full of tanks of all shapes and sizes so it can make wines in hugely varying quantities. Previously, New Hall had something of a mixed reputation, Winward explained: 'Bad packaging and branding, and sweet, heavy wines', though many of the wines the Greenwoods made for other people won awards. Much of New Hall's own wine was sold by the leaseholder system where investors would own vines and then receive a certain number of bottles a year. A friend whose father leased vines still has a sizeable stock of old vintages in the winery's distinctive old black and yellow colours. I've tasted a few bottles of old sparkling wine which, though a bit old-fashioned, have aged surprisingly gracefully. So it's not that the old wines were bad by English standards, it's that the business moved on, through companies like Chapel Down and Nyetimber, and New Hall got left behind.

In 2015, Bill Greenwood's son-in-law Chris Trembath, a local farmer, took over New Hall. Since then he's been trying to modernise the winery via less contract winemaking, a streamlined range and better packaging. In 2021, the winery hardly sold any of its grapes, crushing almost all of them into wine to bottle under its own labels. By English standards, the wines are startlingly cheap, starting at £12 a bottle for the unpretentious and tasty Barons Lane entry-level range.

Just down the road, however, there's a producer with resources that New Hall can only dream of. The winery only began production in 2018, but already it has turned the view of what's possible in England upside down. Danbury Ridge is owned by the Bunker family. Michael Bunker made a lot of money in finance, including a stint at GAM Investment. But unlike nearly every other billionaire with a vision in English wine, his model isn't Champagne but Burgundy. The plan is to make still red wine from pinot noir and still white wine from chardonnay, aiming at the very best of Oregon, New Zealand, Germany – and Burgundy.

It's long been one of the dreams of English winemakers to make a red wine from pinot noir. There's even a protagonist in a mid-nineties John Le Carré novel, *Our Game*, called Tim Cranmer who attempts unsuccessfully to make Burgundy-style wines in England. I think it's meant to illustrate the quixotic nature of Cranmer's character. While pinot noir grows well enough to produce white wine of around 10% ABV, perfect for sparkling wine, such material is not so suited to making a still red wine. The challenge comes when you want to get colour and flavour from the skins. To achieve this you ideally need a warm, dry summer and autumn, something that isn't guaranteed in England, to provide more sugar and less acidity. If the grapes are not ripe, you're going to extract the hard, green notes that you get when biting into an unripe apple.

The winery that has arguably had the most success with pinot noir over the years is Bolney (formerly known as Bookers) in West Sussex. Sam Linter, winemaker and daughter of the founder, has been making reds there since the hot summer of 2003. One year her pinot noir was put head-to-head with some wines from Burgundy on BBC 1's The One Show and came out on top. Bolney was probably the first English red I tried that had proper pinot noir fruit and perfume and, though the acidity was a little pinched, it was clearly a high-quality wine. Since then, other producers in south east England have upped their pinot game.

This is probably most noticeable at Gusbourne in Kent where winemaker Charlie Holland has produced some extremely convincing reds, culminating most recently in a 2018 from the Boot Hill vineyard which was bursting with red-fruit flavours and had a softness and voluptuousness that I hadn't encountered before in an English red. The vineyard is planted with a variety of clones, roughly 50 per cent Champagne and 50 per cent Burgundy. The Champagne ones offer higher yields but don't ripen so fully. It's a question of finding the right sites for the right clones. Fergus Elias from nearby Balfour also has ambitions to make 'Burgundy-style wines in England'. But whereas the Burgundians have had hundreds of years to find the best sites, Balfour was only planted in 2002 and according to Elias, 'We only found our best sites in the last three years.'

Gusbourne began trying to make red wine in 2007, with varying degrees of success. I remember tasting the 2010, which was very much 'good for England'. 2011 was the breakthrough year, says Holland. 'Thirteen per cent natural alcohol. It was a game changer.' It was vital, he explains, to treat vines designed for still wine differently to those designed for sparkling wine, right from the start of the season. This means going for lower yields via green harvesting, cutting bunches of unripe fruit off so that the vine's energies go into ripening the remaining fruit. Sacrificing fruit that could go into sparkling wine that sells for £40 a bottle is a difficult strategy to sell to accountants, but Holland explains that he needs at least 12% of natural alcohol to make a proper red wine. That's when you get 'flavour intensity and colour, and the acidity drops', he says.

English reds are coming of age. In 2022, Balfour released

the 2020 vintage of its Winemaker's Selection Gatehouse Pinot Noir 2020 – a statement wine with a statement price of £60. With its lush dark fruit and spices, it was redolent of something from Austria or Hungary. Perhaps even more impressive though, was Gusbourne's 2019 Pinot, showing what the estate can do in a lesser vintage, creating a delicately English, red fruit-driven style. On my 2022 journey meeting winemakers around the country, I tasted some really rather lovely red wines from 2021, generally considered to be a poor vintage. But with all due respect to the winemakers, these are essentially hobby wines that show what is possible, rather than the more lucrative main business of producing sparkling wine. As Holland pointed out: 'If it's not great for still, we'll make sparkling. If acid is too high and sugar too low [for a still wine], it will be an amazing sparkling wine.'

At Danbury Ridge, however, they have put all their eggs in the still wine basket. Well, almost all – they make a small amount of sparkling wine. The Bunker family had been at Danbury long before they decided, on something of a hunch, to plant vines. Daughter Janine explains: 'We have lived here since 1987 so we knew which were the hottest fields. We weren't completely off our heads.' Their consultant, Master of Wine John Atkinson, retorts: 'They *were* off their heads. If they had planted 15 miles to the west it wouldn't have worked.'

But it turned out that the Bunkers' part of Essex was perfect for vines. Essex has one great advantage over every other wine growing region of England – it's dry. Semi-arid, according to Atkinson. Whereas in Kent and Sussex you might get 70–80ml of rain in September, in Essex it's less than 50ml. The Dengie peninsula, where Danbury is located, has water on three sides which stops the temperature getting too cold, plus there's the warming effect from London, which is only 25 miles away. So whereas in other parts of England you have to harvest the grapes before the rain or fungus gets to them, in Essex they can let the fruit hang and ripen for longer. They also don't have to worry about frost, something that's a massive problem at more inland vineyards.

What really excites Atkinson, however, is the soil. It's mainly heavy, sticky London clay, not dissimilar to the soil near Gusbourne and Balfour. In the past, it was dismissed as unsuitable for quality wine production but according to Atkinson, it's very similar to the soil in Pomerol in Bordeaux. In the sunshine, the clay dries up and holds heat. As Atkinson wrote:[67] 'Even though there are no maps to guide us, intuition tells me that somewhere out there in the "uneventful countryside" between Colchester, Epping and Burnham hides our Pétrus.' The ebullient Atkinson seems unable to contain his enthusiasm about working in Essex. One moment he's discussing the merits of different types of clay and the next he's singing in the style of local punk rocker Ian Dury, or doing a pitch-perfect impression of Mark Reynier, former wine merchant and now a whisky maker in Ireland.

The Bunkers planted in 2014 and took on Atkinson in 2018. Before he joined, he was making wine at a little vineyard in Cambridge: 'In 2016 when I got 10.5% ABV I was punching the air!' he recalls. 'But when I saw what Liam could do with Essex fruit – the extra ripeness and alcohol – I realised I was wasting my time.' Winemaker Liam Idzikowski is a former national hunt jockey from Ireland with some Polish heritage, as you might guess from his name. He graduated from Plumpton in

---

67 timatkin.com

2013 among a class that included some of the most interesting people in English wine: Chris Wilson of Gutter & Stars, Sergio Verrillo of Blackbook, Ben Witchell of Flint, Miguel Symington de Macedo of Rathfinny, and Lucy Winward from New Hall. All of them studied under Australian Tony Milanowski, who now makes wines at Rathfinny.

Idzikowski has done stints at Williams Selyem in Sonoma, which makes a distinctly Californian style of pinot noir, as well as two English wineries, Langham in Dorset and Lyme Bay in Devon. His accent betrays his peripatetic lifestyle, veering between Ireland and the West Country. When he was at Lyme Bay in the calamitous summer of 2012 (the year when Nyetimber made no wine at all) he says you could barely ripen wheat let alone grapes, 'but I saw all these ripe grapes coming in from Essex'. As head winemaker at Lyme Bay he made some very convincing chardonnays and pinot noir from Essex fruit, much of it from Danbury Ridge. He wasn't a fan of everything he made at the time, however. 'I was far from happy with all the wines. And I hated making the mead and fruit cider… even though it was a good learning curve.'

The still wines Idzikowski made from pinot noir caught the attention of the Bunkers. Initially they were planning just to grow and sell grapes, but they were so impressed with the Lyme Bay wines made from their fruit that in 2018 they decided to bottle their own. Those first vintages were made by Idzikowski at Lyme Bay before he moved to Essex full time. The draw was obvious – he was given every winemaker's dream of getting to equip a winery to his own specifications with pretty much unlimited funds, and able to immediately make the next vintage (2019).

Idzikowski is clearly relishing the freedom and budget to

pursue the Bunker dream of making a red Burgundy rival in England. He describes how, previously, the Essex fruit was 'like a workhorse' whereas 'we're now treating [it] like a thoroughbred.' He and the Bunkers work with a vineyard manager called Duncan McNeill who is known as 'Mr Essex' for his intricate knowledge of vine growing in the county. Originally from Yorkshire, McNeill worked at New Hall before setting up his own consultancy company called MVM. He says that the big change he's seen in Essex has been the confidence to leave the fruit out longer to get riper.

Idzikowski's brief from the Bunker family was to make superlative wine. 'Small but premium,' as Janine Bunker puts it. 'We don't want to produce lots of average wine.' They aim to turn out around 100,000–150,000 bottles per year and there are no plans for any tourism offering. There is, though, a purpose-built winery, kitted out with the best-quality equipment and cleverly designed to be naturally cool so there's no need for air-conditioning. According to Atkinson, 'There's no urgency to make a profit. We can take our time. There's no pleasure in making an ordinary wine.' Idzikowski is clearly delighted to have been entrusted with making the wines: 'They didn't bring in a French consultant. They gave me the responsibility.' Any grapes they are not happy with can be sold, and the press wine, from the later pressing of the grapes which contains more impurities, goes to other producers.

The first release I tried from Danbury Ridge was the 2018 pinot noir. Made from a mixture of estate-grown and bought-in fruit from the Crouch Valley, it weighed in at scarcely credible 13.5% natural alcohol. No additional sugar needed here. It was so different from anything from Gusbourne, Bolney or Balfour, tasting more of the New World – Chile perhaps – than England.

It was a deep, dark cherry colour with lots of spice from oak ageing. You can only get that much colour and flavour with very ripe grapes. It was deeply impressive. But Idzikowski says they don't enter their wines into competitions because they taste so un-English that he thinks they won't show well against other English wines.

Danbury Ridge does make a little sparkling wine so in an off year, fruit can always go into this. But so far they've been very lucky, able to make good still wines in 2018, 2019 and 2020. 2021 was touch-and-go with its unseasonably wet July and August. Many growers lost a sizable chunk of their crop but a warm September saved things for Danbury Ridge. As Idzikowski put it: 'We thought we were fucked in August; Duncan said we needed a miracle. But then September was warmer than August.' Even so, they ended up picking some of the grapes in November. I tried a barrel sample of the 2021 from one particular vineyard that was full of pure pinot magic, alive with raspberries, red cherries and orange peel. For Atkinson, this is the thrill of cool-climate ripening. 'When you have just enough light and the right soil, it leaves an imprint on the wine.' The character lies in 'the twists and turns of the grape's journey to ripeness'. In the New World, by contrast, with its more consistent sunshine and climate, you have to pick before the grapes get overripe.

While I was there, we conducted a blind tasting of various pinot noirs from around the world. The Danbury reds are so distinctive that we all guessed them blind, but not because they tasted English. Interestingly, one of the Burgundies in the line-up was so overripe that it didn't really taste like pinot at all. Atkinson and Janine Bunker have taken the Danbury wines to Burgundy, where they met with Olivier Leflaive, one of the region's greats. His response? 'We need to be worried –

the wines are how we used to taste 20 years ago.' As the climate warms up, it's getting harder to make wines with elegance in Burgundy.

Danbury produces a range of pinot noirs including single-vineyard cuvées and relatively cheaper multi-vineyard bottlings which include some bought-in fruit. There are also chardonnays in a ripe, oaky style that also don't taste remotely English. We tried one of the white wines blind within a line-up of wines from all over the world and nobody could guess which one was English. Indeed the Danbury style has come in for some criticism for having too much oak character but having tried various vintages, it seems that after time in bottle, the wines lose the overt oakiness, with the reds keeping the spice and the whites taking on a nutty texture. The entry-level 2020 Danbury Ridge red tasted in 2022 showed beautiful fruit and restrained oak spiciness. We also tried some highly distinctive, experimental sparkling wines made in an oxidised style which tasted a bit like sparkling sherry. They are sweetened with what Atkinson calls 'Essex PX',[68] a mixture of chardonnay and sugar aged in barrel.

Danbury is the most prominent exponent of what's possible in Essex. And with a millionaire's money behind them, they have every opportunity to take advantage of the county's favourable climate. Its success has inspired other Essex growers to bottle their own wines, with Riverview Crouch Valley releasing its first wine in 2022. It's made at Lyme Bay, which has continued its Essex love affair even if Idzikowski has moved on (its 2020 pinot noir won a gold medal at the International Wine Challenge in 2022). With its supplies of high-quality fruit,

---

68 Pedro Ximénez, a Spanish grape variety used to make very sweet wines

Essex is becoming the county of choice for winemakers at the other end of the financial scale from Danbury Ridge, who have to buy in their grapes. And that even means winemakers based in the unlikely surroundings of London . . .

CHAPTER 14

Urban wineries

'I'm happier in the cellar than in a field'

Chris Wilson

The British drinks industry owes a lot to the humble railway arch. Back in the 19th century the arches by stations like London Bridge would be used for storing newly-landed wine and brandy from France, Portugal and Spain. Those by St Pancras were full of beer brought down by rail from Burton-on-Trent. More recently, arches provided a home for wine retailers. In the 1990s it seems that there wasn't an arch in England that didn't house a branch of Oddbins or one independent merchant or another. Laithwaites, now the country's largest online wine retailer, began in a railway arch in Windsor. Not only were such premises cheap to rent but the damp clay brickwork stayed cool, providing a good environment for storing wine. I have fond memories of browsing at Hoults in Leeds while the trains overhead made the bottles rattle.

Railway arches aren't a cheap option anymore, especially as British Rail's successor company Network Rail is selling them off, but these dank, decaying spaces are still providing a leg-up to the wine business. Only now the inhabitants aren't just selling wine, they're making it. Urban wineries are opening right across England's cities, in London, Cambridge and, errr, Tring.[69]

When I first heard about English urban wineries, I thought they sounded like a gimmick. The first, London Cru, was opened in Fulham in 2013 by the wine merchant Roberson. Initially, the wines were made from grapes imported from Italy and France. Why would you do this? Why not just import wine? Perhaps I was missing something. But when the focus shifted to using grapes grown in England, the whole thing made a lot more sense. Most of the best vineyards in England, after all, are within a morning's drive of London. Sergio Verrillo, who runs Blackbook winery, says he can get most of his grapes, from the moment they are picked in the vineyards of the Home Counties into his railway arch in Battersea, within two hours. Compare this with Australia where grapes are trucked across states, sometimes taking days.

The businesses operating from the arches surrounding Blackbook are a mixture of old and new London, with garages and fruit wholesalers nudging up against a charcuterie, a craft vodka distillery and Verrillo's urban winery. And, yes, one of the arches is occupied by a wine merchant, Lea & Sandeman. Oddly, despite its location not far from Clapham Junction – Europe's busiest station, apparently – the air smells decidedly

---

69 Though sadly, Tring Winery closed its doors in December 2022

rural. Verrillo puts this down to the greengrocers washing their produce. It certainly brings the smells of the countryside to London.

With his beard and floppy hair, and that name, you might mistake Verrillo for an Italian artist but he's actually from Connecticut. He originally came to London to work in the music business but gravitated towards restaurants with a stint at Gordon Ramsey's Maze restaurant in 2009. From there he developed a wine obsession that led to working harvests all over the world, with stints in Burgundy, the south of France, New Zealand and, more prosaically, at Greyfriars vineyard in Surrey. Inevitably, he studied at England's best (and only) wine school, Plumpton College, finishing in 2014, and then looked around to see what kind of career was open to him in wine.

Now with a young family, he decided to stay in England for the quality of life but didn't want to get a job with one of the larger firms. 'They're not my cup of tea,' he told me. Having just bought a house in London, he decided to see what he could do in the capital. 'There are so many vineyards nearby,' he reasoned. 'I could bring the English countryside to London.' So in 2017, along with his wife Lynsey, he decided to open the city's second urban winery. Which is where the railway arch came in. When they leased the space it had only been used for storage so they had to install plumbing themselves. And although it wasn't cheap, it was reasonable compared with alternative properties in East London, where many small businesses continue to be pushed out by rising rents.

Right away, Verrillo knew what kind of wines he wanted to make. Not the sparkling wines that are the backbone of the industry, but chardonnay and pinot noir fermented in barrels with wild yeasts and very little manipulation. When I met him, he was running around trying to get wines bottled and had recently broken his thumb so was a touch distracted. But he warmed up as we tasted the wines. It's funny how wine can help people get along. I think it helped that I liked what we tasted so much. Which is just as well, since he wants his wines to compete with the best that the world has to offer, rather than with the local competition.

'I won't drink English wine because it's insular,' he said. 'I want to make really, really good wine. We are a premium brand. I benchmark the wines globally, not against English wine. If you are a big fish in a small pond what does it mean?' It's funny how many winemakers I met in England said something similar – a sign of the increasing ambitions of this country's small industry, perhaps.

Making high-quality wine in London has not been easy. Initially, it was a struggle for Verrillo to obtain fruit. 'No one knew who we were. People were not open to us, not entirely helpful. I would call a vineyard and they would hang up on me.' Sometimes he would buy grapes and 'the fruit would not be good enough. Growers who strive for quality are hard to find. I don't want to deal with you if you are just a cash crop farmer. Wines are only as good as the fruit, only as good as the grower.' He's full of praise for one farmer, Dale Symons in Essex. 'He was the first person to sign up [to be a supplier] and five years on, we have a good, consistent relationship.' Verrillo takes about 8–9 tonnes of grapes from Symons, about a third of Symons' production, which also goes to Chapel Down. 'You have to have an amicable relationship. It has to be symbiotic.'

'It remains a battle when you're competing for grapes with people with far more money than you though,' says Verrillo.

'Some people want 50 per cent more on the previous vintage,' he told me. At the moment, there's too much money chasing too few grapes. With pinot noir selling at £4,000 a tonne, Verrillo has to charge around £25 a bottle to make any money. By English standards, this is not expensive but it is still far more than most people are prepared to pay for a bottle of still wine. And he wants to keep his prices reasonable. At some point, he thinks, with so many new plantings there will be a glut of grapes and then it'll be boom time for the little people, but he doesn't think that will happen for five years at least.

Verrillo could probably charge more for his wines, as they now sell out on allocation. Or 'pompous pretentious bullshit allocation' as he puts it, troubled by the concept. 'We want to be an inclusive not an exclusive brand,' he explains. But in the early days he found selling a problem. 'I had to hand-sell to the trade, utilising all my contacts.' Publicity wasn't easy either. 'I found it really hard to connect with the press.' You'd think a maverick American making very decent wine out of an arch in Battersea would make for a great story, but he had to hire a PR agency in order to get journalists interested.

Whereas some winemakers love the PR side of the business, for Verrillo it's a distraction from what he really wants to be doing. 'Getting it right in the vineyard means less to do in the winery,' he told me. This is something that pretty much all winemakers say these days, but I get the impression that Verrillo really means it. He doesn't add enzymes to the wines and uses minimal sulphur dioxide. With his chardonnay he uses a technique where, rather than protecting the grape juice from oxygen, he allows it to oxidise, which makes the wine less prone to oxidation later in the process. He describes it as like creating antibodies to oxygen. 'Imagine having never been in water before and then you're thrown into an ice bath,' is his way of explaining the shock a wine gets if it has been protected from oxygen and is suddenly exposed to some. With the reds, he includes some whole bunches, stalks and all, with the loose grapes in the pressing process. The idea is to add spice and floral notes but it's a risky move in England when the stalks can unleash astringent, green tannins if not carefully handled.

Unlike other producers trying to work as austerely in England, Verrillo manages to produce consistently appealing wines. There's not a duff one among them. The reds from 2021 – not a great year – were a little heavy on the green stalky notes for me, but the 2020 pinot noir shows the massive potential of Essex fruit. Meanwhile, the whites were uniformly excellent: superb, nutty, barrel-fermented chardonnays; a crisp, saline pinot blanc; and perhaps the most interesting wine of all, a tank sample of his Mix-Up wine from 2021. It's made from two of the most-maligned grape varieties: 40 per cent reichensteiner and 60 per cent müller-thurgau. 'Müller-thurgau is a great variety,' Verrillo argues. The wine is absolutely bone dry with a green apple edge and a delightful creaminess. It's a uniquely English style and because these are unfashionable grapes, he can sell it for less than the other wines, at around £18 a bottle. For me, wines like this are the mainstream future of English wine. Bring on the great grape glut.

Compared with most of those competing for grapes, Verrillo is working on a shoestring budget. He wouldn't go into details about budgets but disclosed that he spent around £100,000 on barrels, tanks, a press and other equipment. Yet Verrillo is a veritable tycoon compared with his former classmate at Plumpton, Chris Wilson, who is making wine in Cambridge at a winery that he calls Gutter & Stars.

Wilson is a former tabloid journalist but was sick of working night shifts and realised that newspapers were in decline. Like many journalists, he'd picked up a taste for wine, so he began writing about it, which is how I got to know him. As fellow wine journalists we used to nod at each other across crowded wine tastings. I was dimly aware that he'd trained at Plumpton when he sent me a bottle of the inaugural wine, a 2020 chardonnay christened 'Daylight upon Magic', made from Essex fruit. I have to admit, I wasn't that keen to try it (writer-turned-winemaker rather set the alarm bells ringing) but eventually I opened it. I was astonished. Not only was it good, it was up there with the best English chardonnays I'd tasted. And while a few years ago such an accolade would have been seen as damning with faint praise, such is the speed with which English wine is changing that is no longer the case. My tasting note captures how impressed I was:

'This is gorgeous stuff. Initially lean and limey, but it fills out with time, bringing lemon and satsuma notes with a noticeable toast and toffee from the oak. Plenty of ageing potential. I'd love to try this again in a couple of years. It might be the best English chardonnay I've had. I love the style – lean but perfectly ripe. Hard to make a comparison, but it's like a more exotic-tasting Chablis.'

The name, Gutter & Stars, from the Oscar Wilde quote 'We are all in the gutter but some of us are looking at the stars', is a reference to Wilson's lack of funds but grand ambitions. He decided against naming it after its home city because, 'As soon as you put Cambridge on the label you age 50 years.' Living in north London at the time, Wilson's training at Plumpton in south-east England wasn't easy, but the gruelling commute was worthwhile, and having classmates Ben Witchell (now of Flint) and Liam Idzikowski (Danbury Ridge) on his doorstep proved a lifeline.

After graduating from Plumpton, he initially moved into wine journalism. The next move should have been to get a job as assistant winemaker at one of the big wineries but Wilson wanted to live in a city. He was inspired by the story of Duncan Savage, a winemaker he had written about. Savage was a winemaker for a large producer in South Africa but got tired of the corporate life so started making wine for his own brand out of a garage in Cape Town from bought-in fruit. He now makes some of the most highly-regarded wines in the country.

So if it could be done in Cape Town why not London or, indeed, Cambridge? Wilson's wife is from Cambridge and he's from Norfolk, which isn't far away. 'If I'd stayed in southern England, I'd probably have got a job in wine and ended up living in the countryside,' he says. His wife, who works for Lego, was hugely supportive of her husband's ambitions and the move from London to open what would still be an urban winery. Wilson found a tiny basement room at the bottom of an old windmill just outside the city centre. The surrounding area has been redeveloped into trendy offices – there's even a vintage Citroën food van – but Wilson's winemaking space is rudimentary. Everything has to go down the steep steps and small doorway, limiting the equipment he can use, which has to be on a Lilliputian scale. 'It affects my winemaking style,' he admits. He uses a mixture of 220-litre fermenters, second-hand barrels from Pouilly Fuissé in Burgundy and a plastic container that looks like a dinosaur egg to make all his wines. He estimates start-up costs at about £10,000, the most expensive items being the pump, press and bottling machine. Space is so limited that he takes the white grapes to Flint in Norfolk, about an hour and a half's drive away, for pressing. It's a proper East Anglian effort, with Essex grapes going from Cambridge to Norfolk and back again.

It's an astonishing story. Wilson moved to Cambridge from London in August 2020 at the height of the pandemic and did most of the renovations on his basement winemaking space himself. His winemaking is rudimentary. He destems his reds using a wire mesh rather than the crushing and destemming machine used by larger producers, and, in wonderfully Dickensian fashion, employs his children to help with *pigéage* (the submerging, by foot, of the skins and seeds from the top of the fermenting red wines to gently release their flavour and colour). At harvest he has friends to help out; a few weeks after my visit, wine journalist Jamie Goode was there helping with bottling.

Just as with Verrillo, Wilson has found it a struggle to get grapes. 'People are trying to charge chardonnay prices for bacchus,' he says. In 2021, one grower promised him a tonne of chardonnay, but only delivered 100kg. 'He won't be working with us again.' At the last minute, another grower lost two-thirds of his crop of bacchus and ortega to fungus. Grape prices have soared from around £1800 a tonne in 2020 to £2100 in 2021. Wilson laments the lack of control he has over the grapes, preferring to literally pick the grapes himself and take the fruit back to Cambridge. Yet he has no plans to own his own vines. 'I'm happier in the cellar than in the field,' he says. As a former freelance journalist, he loves the scrappy, wheeler-dealer aspect of the job, while the same skills have helped with the marketing of the wines too. The winery, the windmill location and the wines themselves are all named after songs – 'I Wanna be Adored' Bacchus is perfectly aimed at all the centrist dads out there – helping create a brand identity for which many bigger and more established wineries would kill.

None of this would matter, of course, if the wines weren't good. The chardonnay, as noted, is excellent. I was less convinced by the 2020 Pinot Noir, very much a 'good for England' sort of wine. But the follow up vintage, the difficult 2021, while coming out pale – more of a rosé than a red – had a haunting perfume, sweet raspberry fruit and not a green note in sight.

Wilson is his own sternest critic. Initially he wasn't keen on the orange bacchus he made, fermented on the skins like a red wine, but he's coming round to it, while he criticised a chardonnay for having too much butterscotch flavour. He sells most of his wines to local retailers and restaurants, delivering by bicycle, or mail order to private customers. He could sell everything to wine merchants but that would be at a wholesale price, meaning less profit, so he does just enough to keep key retailers happy. Talking with Wilson was a fascinating insight into what is required for urban winemaking to be an economically sustainable business. Wilson reckons he can sell up to about 6,000 bottles himself, without the need for an agent or distributor (he was aiming to produce 5,000 bottles in 2022). A retail price of around £20 per bottle and a 40 per cent profit margin means he can make £40,000 a year before expenses.[70] Not bad for a job that for most of the year is part-time and that he can work around his family life. 'I don't employ anyone; I hand-sell all the wines myself.'

While Wilson has no interest in the life of a farmer, Verrillo

---

70 Interestingly, this is almost exactly the same as Tim Phillips from another tiny producer, Charlie Herring, reckons he makes from farming his one-acre walled garden in Hampshire. So it seems that you can make a living from wine as long as you are nimble and your wines are good enough

would at some point like to own his own vineyard. His big dream is to have a large-scale winery, restaurant and tasting room – in Chelsea. Maybe one day. For the moment, neither Blackbook nor Gutter & Stars are regularly open to the public, something that would mean taking on staff. But part of the business model of an urban winery is the direct proximity to customers. The original London Cru offers a package that includes a tour and bottles of wine, while Renegade, another urban winery in East London, has a tasting room and bar. Tourism is an increasingly important element of most English wineries' offering – and some do it on a very grand scale indeed . . .

CHAPTER 15

Tourist attractions

'When I go to Tillingham, I always end up bumping into people I know. It's like going to Shoreditch House'

Hannah Crosbie

U ndoubtedly the strangest experience on my journey around southern England was when I found myself in a converted oasthouse[71] in Sussex with a Tibetan singing bowl on my head. These bowls are made from tuned metal and when tapped with a hammer produce an eerie noise intended to aid meditation. I found it not an altogether unpleasant experience, though I did feel a little like Michael Caine in *The Ipcress File* when he's undergoing a sonic assault designed to turn him into an automaton.

---

71 Distinctive conical-shaped brick buildings that were originally built for drying hops

My hosts were winemaking couple Nick and America Brewer. He's English, she's very, very Brazilian and owes her magnificent name to being born on the 4th of July. They weren't trying to brainwash me, of course. To be honest, I'm not entirely sure what they were trying to do, but the bowl incident was just one of several surreal elements of my visit to Oastbrook vineyard, near Bodiam Castle on the Kent/ Sussex border. On arrival, guests are greeted with a rustic sign pointing them to the winery and a 'Hobbit House'. Yes, the Brewers have a small house on site that guests can stay in, complete with round windows and grass on the roof (although to avoid copyright infringement they had to rename it 'Halfling', America having run a competition to find an alternative name that still nodded to Tolkien).

With their Tolkien house, Eastern spirituality and colourful Instagram presence, you could be forgiven for not taking the Brewers seriously. At the onsite laboratory they put on lab coats and posed for pictures as if they were big kids pretending to be winemakers. But they are deadly serious about their wine. Nick recounted proudly how America was top of her year in plant biology during her BSc in viticulture and oenology at Plumpton College; he did the one-year business-focused course but doesn't just focus on the commercial side of things – he makes the wines, too.

Those wines have won multiple awards and received plaudits from such renowned critics as Matthew Jukes. The Oastbook still whites had not yet been bottled when I visited, but a pinot meunier, a pinot blanc and a chardonnay were some of the most elegant I tasted on my travels across the country. Production is tiny – less than 10,000 bottles a year in total, with some lines made in miniscule quantities. If it was based in California, Oastbrook would be a cult winery selling strictly on allocation for $100 a bottle. But as it's England, the excellent Oastbrook Chardonnay 2021 sells for £27.

Like many in English wine, Nick Brewer has a corporate background, having had stints working in asset management and the energy sector. He doesn't have the kind of wealth of Mark Driver at Rathfinny or Eric Heerema at Nyetimber, however. Brewer estimates he's put about £1.5m into Oastbrook, which they planted on an old hop farm that used to supply Guinness. They had to see a return right from the start, and as a result, Brewer describes the business as 'highly diversified'. As well as accommodation on site, there's a large hospitality space and a professional standard kitchen. In the summer, they put on such ambitious musical events as Benjamin Britten's opera *A Midsummer Night's Dream*, all presided over by the one-woman publicity whirlwind America Brewer.

America sees Sussex, despite its proximity to London, as relatively undiscovered in terms of its tourist potential: 'As a Brazilian I see bigger than you guys do. People don't appreciate how beautiful Sussex is.' Winemaking at Oastbrook would not be possible without tourism, which not only attracts paying guests to the winery, but allows for direct sales of bottles without a wine merchant taking a cut. As a result, they sell everything they make from the cellar door, their website and via local restaurants. With such a set-up, they can produce up to 50,000 bottles a year. Anything more than that would require much wider distribution, larger facilities and more employees. Nick Brewer describes those next-level businesses, producing between 50,000 and 200,000 bottles per year, as 'neither fish nor fowl', meaning that, in his view, you either want to be big or very small. And if you want to be small, then you have to build tourism into your business model.

Clare Holton at Brissenden in Kent also finds tourism invaluable. Unlike most people who work in wine who were following their own dream, she was following that of her husband, who had always wanted to own a vineyard. In 2017, in partnership with her father, that dream became a reality. Sadly, the marriage didn't last, but the vineyard did, and rather than sell up, Holton and her father persisted and have managed to turn it into a viable business. They have seven acres of vines and sell 90 per cent of their crop to the nearby winery Balfour. Brissenden's grapes are highly prized and previous owner David Rackham had always sold them to Owen Elias, initially at Chapel Down and then at Balfour when Elias moved. According to Fergus, Owen's son and successor, Brissenden's old block pinot noir is 'one of the most interesting plots in the UK'. In a good year like 2018, the Holtons get around 38 tonnes of grapes (around 5.4 tonnes per hectare), making it commercially viable. But in a horror show like 2021, they struggled to get four tonnes in total.

Regardless of the size of the harvest, Holton keeps back around 10 per cent of the crop, which South African winemaker Kobus Louw, who works with a number of other local vineyards, turns into a still bacchus and a sparkling blanc de noirs. This works out at around 700 bottles of each wine per year, which Brissenden sells locally, direct to customers. A large part of this comes through cellar door sales. 'We don't have any economies of scale,' says Holton. 'I can't really have a wholesale price or I wouldn't make any money. I can't afford a middle man.' To bring people in, she runs tours of the winery at £25 per customer. Being so small means she has to think about every bottle she opens, so will only run tours for six visitors or more (and there are definitely no free samples for journalists). At the end of the tour, most people buy a bottle, but not all. 'Some people are not mad on wine, they've just enjoyed the day out. It's the experience,' she says. 'They've already paid, they've had a lovely time, and I've earned enough money not to be out of pocket.'

Visitors are Brissenden's lifeblood. They are attracted through word of mouth; I heard about the winery through a friend whose children go to school with Holton's children. Facebook is particularly helpful too, though Holton is not convinced by Instagram. One reel she made got 46,000 views, but she doubts it sold a single bottle of wine. There's also a small campsite among the vines with space for four tents, and during the pandemic Holton ran Halloween trails for children, though she admits that they didn't really make any money. It's all evidently a huge amount of work making ends meet but Holton is indefatigable and clearly loves the life. She can count on the local community for support – her pickers are locals – and she praises Rackham and the Eliases for their support.

While Brissenden pulls in local families, Tillingham in Sussex is casting its net considerably wider. As wine communicator Hannah Crosbie wrote in *The Times*: 'When I go to Tillingham, I always end up bumping into people I know. It's like going to Shoreditch House.' Founder Ben Walgate explains: 'From early on we had celebrities and people in fashion wanting to book parties. It became a bit of a hangout. I'd walk out in the afternoon and there would be Lily Cole and Paloma Faith having a pizza on the grass.' The Tillingham experience is very much not for everyone; we have friends who visited who thought they were treated with disdain by the staff. But it does appeal to trendy urbanites who like the idea of being in the country, seeing picturesque pigs and tasting wine, but don't want it to feel too different to London. As with Oastbrook,

Walgate sees himself as a winemaker first and foremost, rather than a hotelier, but recognises that unless you're a billionaire, you need the visitors to pay the bills.

While some of the money men don't do tourism – neither Nyetimber nor Danbury Ridge is regularly open to visitors – Mark Driver at Rathfinny considers it an integral part of the business. There's a 35-cover restaurant on site called the Tasting Room whose chef, Chris Bailey, has worked in swanky restaurants in London and Madrid. There's also a small hotel. The mark-up at the restaurant on their own wines is a very reasonable 30 per cent (100 or 200 per cent is normal in London). They also sell still wines and spirits that aren't available anywhere else. Rathfinny attracts around 40,000 visitors a year who come to look round, eat and taste the wines. Most of the larger wineries are investing heavily in food as well as their wines (Wiston, also in Sussex, has just opened a restaurant called Chalk) but in a rural setting, the challenge is finding and keeping trained staff.

Leaving Rathfinny, I saw a huge neon sign with the words 'A Place Beyond Belief' facing out across the South Downs. At first I thought this was a bit of corporate hubris but it's actually an artwork by Turner Prize-shortlisted British artist Nathan Coley. His speciality seems to be big neon signs with enigmatic slogans. There are six such artworks dotted around Sussex, commissioned by an organisation called Sussex Modern which brings together wineries, artists and restaurants to promote tourism in the county.

Sussex Modern is not the only body trying to market the county as a wine destination. I went on a press trip with another organisation called the Great Sussex Way. Our guide was a smartly-dressed young man called Lawrence Abel who is also the aide-de-camp for the local MP. Abel says he wants to see

Sussex wines at events 'instead of Veuve Clicquot'. At the restaurant where we had lunch, although all the food and beer was local, they didn't have a single English (let alone Sussex) wine on the list. Part of Mark Driver's vision for a PDO[72] (protected designation of origin) for Sussex is to drive the distinctiveness of the county's wines. I ended up having a series of rather roundabout conversations with Abel and the other writers on the trip about what was distinctive about Sussex compared with say Kent or Surrey. The truth is that, compared with Yorkshire or Cornwall for example, Sussex doesn't really have a strong identity. Nevertheless, other south-eastern counties are also trying to promote regional identity, as with the Wine Garden of England in Kent or Vineyards of Hampshire.

At Hambledon in Hampshire, the visitor side is looked after by Caron Fanshawe who previously worked with wine tourism company Arblaster and Clark, which runs vineyard tours across Europe. 'It never occurred to me in the early 2000s that I would be doing this in England,' she says. But with many vineyards in south-east England less than an hour's journey from London, it makes sense. Those that are further afield tend to be in beautiful parts of the country already popular with visitors, such as Cornwall, Devon and Dorset. During the summer months, Camel Valley in Cornwall is usually booked up for tours days in advance. Sometimes they have to turn people away. Then again, owner Bob Lindo has invested a lot of time and effort in turning Camel Valley into a destination. I noticed that he replies to every review on Trip Advisor, with an apology and explanation for the rare bad one. It's this personal touch that people love. 'We don't have departments,' he says. 'Six of us do everything.'

---

72 See 'Money Men', chapter 6

Lindo sells about 50 per cent of his production direct to customers. He could sell 100 per cent, he says, but feels that, for the reputation of the winery, it's important to be visible in the likes of Fortnum & Mason and top restaurants.

The Covid overseas travel restrictions imposed in 2020 and 2021 saw UK visitor numbers to wineries explode. As Mark Dixon of Rathfinny says: 'In some ways, Covid was brilliant for English wineries.' Though the restaurant business disappeared, wineries more than made up for it with online and visitor sales. Such was the deluge of visitors that Biddenden in Kent, which had up till this point run free tours as a way of selling wine, began charging. 'We should have done it years ago,' says Tom Barnes, whose grandfather founded the estate. At Langham in Dorset, the outdoor undercover area became a community hub with retirees dropping in for tea.

But Robin Hutson, the hotelier behind The Pig group, is not yet sure that English wineries are a big draw for tourists. Referring to his own guests, he says: 'I'm not entirely convinced people are coming to taste the wines. It feels like an additional activity when they are here.' He thinks that this is beginning to change, though. The inspiration for the South Downs branch of The Pig, with its vineyard out front, was Australia, 'where you're in a restaurant overlooking the vines'.

He's not alone in being inspired by New World vineyard tourism. Art Tukker, owner of Tinwood in Sussex, says: 'I worked out in New Zealand and saw how tourism works in the vines.' Originally his family grew lettuces for supermarkets, but tired of being beaten down by supermarket buyers, so decided to diversify into grapes. Tukker's father was also a major investor in Ridgeview. Today, about 75 per cent of his crop goes into Ridgeview own-label wines with the rest coming back to him as

Tinwood Estate wines. Production of Tinwood wines is about 50-60,000 bottles per year, with the majority sold direct to visitors. When I visited Tinwood on a warm May afternoon, the terrace was packed with customers enjoying the estate's wines with cheese platters. The setting is all decked out in the slick wood-and-glass aesthetic of international vineyard tourism. According to Lawrence Abel, the young people of Chichester consider it 'quite fancy' to go for an evening at Tinwood.

While Tukker is inspired by New Zealand, Hundred Hills near Henley-on-Thames is following the high-end Californian model. The Oxfordshire estate was founded by Stephen and Fiona Duckett in 2014. Just as with prestigious Napa Valley wineries, there's a club you can join to receive priority access to the wines and a chance to visit the estate, which is not generally open to the public. The £600 cost includes 12 bottles of wine, a discount on future purchases and a visit for six friends. For a time, Hundred Hills offered an even more exclusive £20,000 Hundred Club, an investment scheme whereby members received £1000 worth of wine every year for five years plus the opportunity to hold events at the winery – and they got their money back at the end! I was all ready to sign up but sadly the scheme has served its purpose and is being run down. For a more modest £500 a year, Gusbourne in Kent offers a membership called Gusbourne Reserved. For your fee you receive 12 bottles of sparkling wine a year plus discounts and membership events. Considering the wines you'll receive range between £40–60 a bottle, it's rather good value.

Good value is, of course, relative. Whether your tastes run to red trousers and branded gilets or nose rings and tattoos, a day at an English vineyard is usually going to be expensive. Tillingham might have a 'hipster' image but a 50cl bottle of its still pinot meunier will set you back £60 in the on-site bar. English wine is never going to be cheap. In the past, vineyards would let people look around for free in the hope they would buy some wine in the shop, but as Frazer Thompson, former CEO of Chapel Down, says, 'You'd get coach parties who'd come in and buy a jar of jam.' Nowadays, it's quite the opposite, with the aim to extract as much money from tourists as possible. At a Wine Garden of England event at Squerryes Hall in Kent, there were mutterings from visitors about the price of the food. Sensibly, we had smuggled in a picnic.

Despite the cost, there seem to be plenty of people prepared to pay for a day out in the vines. It's vineyard tourism with a decidedly British twist. Chapel Down was packed with tipsy, slightly rowdy visitors when I visited; in 2021, over 55,000 people came through its doors. Denbies in Surrey even has a little train to take you round the vineyard. Well, sort of. Sadly it's not a real train, just a car pulled by a Land Rover. Wouldn't one of those little steam trains driven by enthusiasts in blue peaked caps be more fitting? Certainly something for the children to do while Mummy and Daddy 'taste' wine.

One vineyard owner in particular has huge ambitions for visitors. Yes, it's our old friend Mark Dixon of Harlot fame.[73] His plan was to build a £30-million, Norman Foster-designed English wine experience called English Wine Vaults at Cuxton on the River Medway that would not only be a tourist attraction but also an underground winery where Dixon's various wines would be made. His ambitions were to produce 4–5 million bottles a year and receive around 65,000 visitors in a place that, from the architect's plans, looked decidedly Teletubbies-inspired, with its

---

73 See 'Fizz Wars', chapter 8

green roof and vast size. Despite this grand vision and some support from the local council, the plan was rejected following local objections. Though knowing Dixon, I'm guessing he hasn't given up on the idea.

There's been much gleeful rubbing of hands within English wine at Dixon's travails. But the Medway towns of Kent like Chatham and Gillingham could really do with a boost to the local economy, which has still not fully recovered from the closure of the Royal Navy dockyards in 1984. It's just a shame to me that the proposed architecture for English Wine Vaults is so generic – the plans for the winery make it look like a European art gallery or a Chinese distillery. This lack of individuality plagues much English winery architecture. Perhaps because the industry is so new here and therefore heavily influenced by developments in South Africa and New Zealand, many owners tend to follow their style rather than taking a fresh approach. There's usually lots of weather-proofed wood on the outside, along with glass walls and terraces looking out onto the vineyard. The feel is Michelin-starred restaurant but with vines. It's all a bit bland. Sometimes seeing another achingly tasteful restaurant made me long for the faux-French châteaux that some Chinese wine producers build to lure in tourists. You can see why the Brewers went for a Hobbit – sorry Halfling – house.

Maybe because I'm a wine writer, what I love in a winery is the history, the old equipment, the characterful buildings. But of course there's not much of that in English wine. Henry Laithwaite thinks that it will come with time – it is, after all, a very young industry. He also isn't a great fan of the rather corporate visitor experiences at many English wineries. 'When I visit a winery, I like to sit on a bucket in the cellar and taste wines.' Funnily enough, the bucket experience isn't one that most wineries charging £25 a pop tend to offer. Laithwaite does welcome visitors personally, as he understands that it is vital to build up a mailing list, but I get the impression he'd be much happier out in the vines. 'One thing I learned from my father [wine merchant Tony Laithwaite] is how important it is to know your customers.'

Some of the best visitor experiences I have had have been at the longer established wineries. These might not be the most glamorous, get the column inches or even make the best wines, but they give visitors a sense of place. With its grand Victorian main building – built by the railway company – housing the winery, and old railway sleepers supporting the vines, New Hall in Essex is particularly attractive. It's a similar story at Biddenden, whose location is advertised by a wonderfully old-fashioned signpost a few miles down the road. On site there are some traditional old fibreglass tanks, used to store cider, while the wine and cider are made in old farm buildings. Langham in Dorset is another winery that has the feel of a working farm rather than a rich man's showpiece. There's a sense of authenticity and history to these places, in contrast to the *Grand Designs* showiness that's a feature of the more contemporary vineyard architecture. It was great to visit Hambledon, the country's oldest winery, in Hampshire, and see that they were building a new visitor centre with oak beams made by local craftsmen. A bit of vernacular works wonders.

The person who seems to have grasped this best is Stephen Duckett at Hundred Hills in Oxfordshire. He might have been inspired by California, but the wine is made in what Duckett describes as a 'traditional Chiltern barn'. It was designed by local architectural practice Nicholls, Brown and Webber and was judged 'building of the year' by the Chiltern Society. With its black stained wood and brickwork, it sits perfectly in the surroundings. It's even better close-up, with beautifully detailed woodwork which is now home to some rare long-eared bats.

'We're trying to make English wine, so the winery has to be of its place,' says Duckett. He adds that while South Oxfordshire County Council is notoriously tough on planning permission, it was so impressed with the design that it sailed through the committees. Rather as the Gothic styling of a Victorian railway station conceals its industrial purpose, so the traditional exterior of the Chiltern barn houses a contemporary interior, but without feeling like a pastiche. It helps that the barn is nestled in a valley where vines grow up the steep hillsides. On an early autumn day with the leaves just changing to yellow and orange, it's up there with the most beautiful vineyards in the country, if not the world. Currently it's only available for tours for members of the Hundred Hills club, though you can hire it for weddings and other large events.

The only place that comes close to Hundred Hills for sheer beauty is Breaky Bottom in Sussex – a winery that is similarly hard to visit, though for different reasons. First you have to negotiate an unmade road, which I was worried would do all kinds of awful things to my old Mercedes. It's worth the journey, however, to see the squat cottage perfectly nestled in a steep, vine-covered valley. Best of all, there's history here, ancient history by English wine standards, and stories, and wine that tastes like it couldn't come from anywhere else. The family winery is not really set up for tourism, however. As Peter Hall succinctly puts it: 'People say we should open for tourism. We don't want to!' You can send them an email and if they have time they might show you around. It's not that Peter and Christina Hall are averse to having people visit. They have

CHAPTER 16

Reaching the customer

'The only question that matters is who's going to buy your wine'

Charles Simpson

'It used to be people who wanted to be like Russell Crowe in *A Good Year*. Now it's people with a plan for how to sell a brand.'[74] That's Ed Mansel Lewis from estate agent Knight Frank on the individuals now moving into English viticulture. And perhaps best epitomising the not-Russell Crowe tendency are Ruth and Charles Simpson at their eponymous vineyard in Kent. They're a British couple, Ruth coming from the Grant family of Glenfiddich fame, but they learned how to make and sell wine in southern France.

Rather than the often romantic types that populate wine-making globally, the Simpsons brought a hard-nosed business

---

74 The *Financial Times*

acumen to vine growing. At one point in the late nineties, they looked at buying vineyards in Central Otago in New Zealand and Margaret River in Australia but, having outlined a formula for the price of a bottle based on the cost of land per hectare, the cost of cultivation for the first three years and the likely yield per hectare, they concluded that the land in those two regions was just too expensive. 'We'd missed the gold rush by four years,' says Charles.

Instead, the Simpsons went to the Languedoc, where land was cheap. In 2002 they bought an estate that they named Domaine Saint Rose which was already producing wine, meaning they could start making money from the beginning. James Herrick, an Englishman who had a Languedoc wine brand in the nineties which he sold to Australian beer group Foster's, gave them some advice: 'Only you two can sell the brand. Sell! Sell! Sell! Leave the wine up to others.' So they hired a winemaker and concentrated on sales and marketing. They made themselves popular locally, since the majority of their business was export so they weren't competing with their neighbours. But they had an advantage over other Languedoc producers as they spoke English and, says Ruth, 'We did what we said we would do. We answered the phone and email. Most local winemakers were either out in the vineyards or at the feria [the annual local festival].'

'The only question that matters is who is going to buy your wine,' says Charles. And with its strong branding, Domaine Sainte Rose proved very successful, a stalwart in Majestic and Waitrose. The Simpsons, though, were frustrated by the low prices. They had hoped that they would be able to change people's perceptions of cheap Languedoc wines but never cracked the £15-a-bottle barrier. 'Languedoc has amazing climate and potential, but to change the reputation will take generations,' says Ruth. That said, other Languedoc estates like Grange des Pères and Mas de Daumas Gassac sell their wines for far more than £15 a bottle, so perhaps the Simpsons were missing a trick.

In 2011 they decided to move back to Britain, while retaining ownership of Sainte Rose. They saw an English vineyard advertised in *Decanter* magazine and ended up buying 30 acres in Kent. 'The land was cheap as chips,' Charles recalls. They had no links to the county but thought it was the best place to grow grapes in England. 'East is best,' he continues. 'There's a reason why Kent is known as the garden of England.' English wine expert Stephen Skelton was engaged as a consultant but the Simpsons brought their own expertise to the venture, including hiring planting crews from France at half the price of the German teams that everyone else in England used. They had already made 'tons of mistakes' in France, so knew what not to do. Initially they planned just to make sparkling wine but realised there was a gap in the market for good-quality still wines, so planted some non-sparkling clones[75] too. Ruth Simpson admits that, 'there were no bigger cynics of [English] red wine than us. But there was huge demand, and not just nationally. Plus it's a great way to get our name out sooner, before the sparkling wine is ready.'

The vines went into the ground in 2014, and rather than having to build anticipation without any product, like Rathfinny did, by 2016 the Simpsons had a small batch of still English wine to sell. Then with the bumper 2018 vintage, they had a range of wines, including an unusually high-quality red. Simpsons Wine Estate seemed to emerge fully formed overnight.

---

75 See 'Grape Expectations', chapter 12

But behind the scenes, there had been an immense amount of thought, work and money put into it to achieve such a profile. Most importantly of all, they got the branding right. The bottles looked different from anything else on offer from English wineries. Local pubs sported Simpsons-branded umbrellas. They even got their sign up at Canterbury East railway station. And just as in France, they left the winemaking to someone else while they went about publicising the product. Talking to the two of them, I had the sense of a finely honed double act. Even when they interrupted each other, it felt like it was choreographed. These people are slick. In an English wine scene where many brands seem to morph into each other, the Simpsons stand out.

What the Simpsons did was the opposite to how most people approach wine, which is to start with the wine and then work out how to sell it. According to Ruth, this is the key thing most people get wrong. 'The worst goal is to make the best wine in the world,' she says. Two people who might disagree with this are another husband-and-wife team, Stephen and Fiona Duckett at Hundred Hills. Working closely with the late Michel Salgues of Champagne Roederer, they planted vines in south Oxfordshire in 2014 with the simple aim of making the best sparkling wines in England. And yet their philosophy is not so different to that of the Simpsons. Stephen is no dreamy romantic. He background is in internet sales and marketing and he says that when starting a business, 'the very first question you ask is "how are customers going to buy and why?"'

'Wine is the most crowded field in the world,' he adds, so he and his wife knew they had to make theirs stand out. Whereas the first and second wave of English fizz producers were gunning for grand-marque Champagnes like Pol Roger or Veuve Clicquot, Hundred Hills worked far more like a small

Burgundy domaine. Production is modest, at around 50,000 bottles, and it has no ambition to make more. There is no non-vintage release; the wines that came on the market in 2022 are all vintage cuvées made up of different parcels of grapes from the estate. The beautifully-designed back labels show a map of exactly where the batches in each bottle came from, meaning they have something different to talk about every year. 'I don't know how I'd sell non-vintage every year,' Stephen says. 'There's too many people doing the Champagne house model of pile it high and sell it cheap. It's not a good model for the 21st century.'

Duckett, who is the spitting image of English actor and comedian Stephen Mangan, admits that it was hard to take on the hegemony of Champagne with its 'hundreds of years of marketing'. First, he says, you need confidence that your product is good enough. Then his sales model is based around what he calls 'accelerated word of mouth'. He elaborates: 'Talk to people one-to-one as often as you can. Then do that with scale.' This involves selling wine directly to customers through the Hundred Hills wine club. Being a member is the only way people can visit the estate, a business model which builds up a loyal band of club members who will spread the word. The other major sales avenue for Hundred Hills is through high-end restaurants and clubs. The Ducketts are working with restaurant consultant Douglas Blyde to get Hundred Hills in front of sommeliers. (And not just sommeliers. Blyde was furious with me when, in a panic about getting the book finished on time, I cancelled a visit to Hundred Hills.) The location of the vineyard helps with this; it's close to vaunted restaurants such as Le Manoir aux Quat'Saisons near Oxford and The Fat Duck at Bray. Indeed the Ducketts employ a former sommelier from Le Manoir, Rupert

Crick, as hospitality manager. And of course the London market is on its doorstep. At the moment, it is listed in around 40 Michelin-starred restaurants.

Another benefit to the location is the proximity to Oxford University colleges, some of which have a lot of money to spend on wine. Hundred Hills produces a special label for Christ Church, which has taken 10,000 bottles to celebrate its 500th anniversary. We didn't get that at Leeds University. Everything from the wines themselves to the branding to the marketing has been very carefully thought through. I was surprised, therefore, given all the couple's talk of the importance of exclusivity, at how affordable the wines are, with the entry level Preamble coming in at under £40 and the premium cuvées at £60. Of course, affordability is a relative concept. One place you will never see Hundred Hills, however, is in supermarkets. According to Duckett, high-end restaurants won't take your wines if they are also sold in supermarkets.

Hundred Hills' prices mirror those of another brand that is successfully going the upmarket route, though on a much bigger scale[76] – Gusbourne. But it's an increasingly crowded market. According to Charlie Holland, chief winemaker and CEO, every new sparkling wine brand wants to get into fancy restaurants. 'Sommeliers love Gusbourne but there's a lot of competition for space in high-end dining.' He's had some 'scary conversations' with new producers whom he asked 'Where are you going to sell it?' Many newer producers won't last, thinks Frazer Thompson, former CEO of Chapel Down. His view is the

---

76 Rather like Dom Pérignon, Gusbourne won't reveal its annual production, but on a 93-hectare site, I'd estimate they can probably make around 400,000 bottles in a reasonable year

industry will go through a period of consolidation, with bigger producers gobbling up smaller ones.

The dream listing for any English wine producer is at one of The Pig hotels across Southern England. Each hotel stocks around 30 English wines. In 2014, owner Robin Hutson took a big leap of faith in English Sparkling Wine by, as he puts it, 'suppressing Champagne by-the-glass in favour of English fizz'. The move required a high level of staff training so that they could confidently sell a lesser-known wine. Hutson does offer Bollinger for those who remain unconvinced but says that almost all customers are delighted with the English offering. As a result, he estimates that he sells around 40,000 bottles of English wine per year across eight hotels. The Sussex branch near Brighton even has its own two-acre vineyard on site. It's not just for show either. The vines planted in 2019 should bear fruit in 2023 and none other than ex-Wiston, ex-Nyetimber man Dermot Sugrue will make the wines at Artelium winery near Brighton. Hutson has also made what he describes as a substantial investment in Dermot and Ana Sugrue's own brand, Sugrue South Downs. Rather than buy up land and then find a winemaker, Hutson is betting on the Sugrues. He describes Dermot as a 'huge talent' who had, up till this point, been 'underfunded'.

Underfunded certainly isn't how you would describe Gusbourne. Visiting the estate I was struck by the amount of gold livery on show. The ice buckets have gold-coloured linings and the branded pens are gold (not real gold, sadly). The team has just launched Fifty One Degrees North, now England's most expensive wine. Charlie Holland is a naturally modest man and looked a little sheepish when I asked him about the price. Apparently they did extensive testing against Champagne counterparts like Krug and Cristal before coming up with the

£195 price tag. Nyetimber too is seeking to ape the luxury feel of grand-marque Champagne; it has recently taken on Richard Carter, formerly of Rolls-Royce, to burnish the brand's profile. According to owner Eric Heerema: 'Richard coming from Rolls-Royce brings tremendous ultra-luxury brand experience. We're a very passionate wine producer but also a British luxury brand.' Carter was made director of Nyetimber in July 2022 but resigned this position in August so it's not entirely clear what his role is now.

While aiming at the luxury end of the market might work for Nyetimber, the problem for rivals is, as Sergio Verrillo from Blackbook winery puts it, 'Everyone is trying to be [Château] Margaux[77],' via 'ambitious pricing'. 'We can't all be porn stars,' he says at one point. I think I know what he means. It does seem the image and the pricing of some wines has got ahead of quality in some places. Furthermore, with so many people chasing the same image, there can be a sameness about English wine. From the outside, the industry appears – at least to me – very corporate, which is odd as it's such a young industry populated largely by small start-ups and family enterprises. Perhaps it's because so many producers come from a finance background. Or it might have something to do with proximity to marketing and PR agencies in London who are all preaching the same thing. It was refreshing talking to Surinder Bassi from Meopham Valley Vineyard, a small Kent brand, who is selling his wine in local nightclubs (he provides branded cushions for VIP areas) where it's apparently going down a storm.

As well as presenting 'The Wine Show' on TV, Joe Fattorini

---

77 One of the five 'first growths' of Bordeaux along with Châteaux Latour, Lafite, Mouton-Rothschild and Haut-Brion

also consults in the wine business. He told me about his experiences with one particular English winery. 'They are hypersensitive to the idea that doing something different will make them look foolish. So all ideas come back to a slightly different colour palette on an identikit campaign that says "Brand X is the Burberry/Smythson's/Range Rover/Mulberry of English Sparkling Wine".' At the moment, this tactic seems to be working – at least for some – but Fattorini foresees trouble further down the line: 'Luxury is what writer Luke Burgis calls "a thin desire". It lacks moral imagination. "Thick desires" are rooted in something real. They're built up over time, like layer upon layer of strong rock that sits under the surface of a pile of leaves. They have a history and continuity.'

Henry Laithwaite at Harrow & Hope admits that the English wine world is a 'very PR-led industry' but thinks that distinctiveness will come with time. 'We have to go through a few years of growing grapes and try to build the tradition,' he says. Christian Seely from Coates & Seely doesn't think the multiplicity of brands making sparkling wines is a problem as long as they are distinctive. 'I don't think they resemble each other very much. I taste all of them every year at least once. There are striking differences in personality, between the ideas behind them and the places they come from.' He compares England to Burgundy, where there are dozens of producers making wines that might appear very similar but, on tasting, have distinct personalities. 'We are just beginning to see that.'

One of the things I found so surprising researching this book is that brands that might appear boring from the outside actually have rich stories and interesting people behind them. I'm not one to tell Nyetimber how to market itself but it has arguably the best story in English wine. Yet you'll struggle to find any mention of the people behind that story, Stuart and Sandy Moss, on its website. Instead, the branding is all about the Domesday Book, which has nothing to do with the wine that Nyetimber makes today. Compare that to Veuve Clicquot's communications, at least to journalists, which makes much of the legendary widow herself. Frazer Thompson thinks that English wine producers are largely hopeless at marketing their products: 'We haven't learned how to tell stories. Veuve Clicquot, the widow, became a brand. Who is becoming a brand in English wine?'

The other trick English wine producers are missing in their drive to be the new Champagne is that wine is an agricultural product. People like their cheeses and ciders marketed with rural imagery, so why not wine? Stefano Cuomo from Macknade food hall in Faversham says that the pendulum is swinging away from the slick marketing of the City boys. He cites Biddenden, with its strong family image of Julian Barnes and his sons Tom and Sam, and their classic ortega with a label that hasn't changed much since the 1970s. 'They're rooted in the countryside,' he adds.

Hattingley Valley in Hampshire is also presenting a fun face to the world, or at least that's how I see it. Much to the horror of some in the English wine community, the brand has gone all-out with British branding. Justin Keay wrote in wine industry website *The Buyer*: 'Hattingley's bottle neck collars even declare: "Proudly Made in Great Britain." To this rattled and disillusioned metropolitan Liberal who has spent much of the past three years apologising for being British, it all seems a bit much.' It does look rather Austin Powers but it stands out, and it's fun. I sometimes think of the vigour with which alcoholic drinks were sold in the past and wonder where the sense of joy has gone.

Think of those jolly Belle Époque Champagne adverts or the amazing Cubist and pop art Campari designs. There's nothing of remotely that quality in the branding of English wine; its advertising seems to have forgotten how to entertain.

Of course, it's not easy taking on Champagne head-to-head. Champagne has had decades of focused marketing to the extent that it isn't really seen as a wine, but its own category. British customers who are reluctant to spend more than £6 on a bottle of wine will happily spend £40 on a bottle of Veuve Clicquot. When you think of Champagne, you think of Formula 1, Marilyn Monroe and James Bond. It's hard to compete with that. As Fattorini puts it: 'England still sells sparkling "wine". Champagne sells something deeper.'

Elisha Cannon from new wine brand Folc laughed when I asked whether the English wine industry was good at marketing. 'I think we can do better. Wine as an industry is quite traditional. The way it markets itself assumes that the consumer is the same person it was 20–30 years ago. They aren't. Are we trying to attract sommeliers or are we trying to reach Mr & Mrs Smith?' Her idea, with her husband, was a simple one – make good English rosé and market it well. The pair had noted that, despite rosé being the only category in wine that is growing, and Britain being the second largest market for Provence rosé in the world, there were very few people producing an English version. Thus Folc was born. It's made by Defined Wine in Canterbury from grapes grown in Kent, Sussex and Essex. Obviously, it doesn't taste much like the Provence original, but it's the right colour and comes in a fancy bottle, which is half the battle in appealing to consumers.

With such wines, social media is increasingly important in spreading the word. If people want to find out about wine,

Cannon says, 'It's not Google, it's Instagram or Tik Tok.' She partners with brands such as hotels or influencers with a similar outlook and image. The brand's Instagram account (I'm too old for Tik Tok) shows an idealised rural image with Cannon and other photogenic types enjoying Folc. It's all very aspirational and has the same aesthetic as hundreds of influencer and wannabe influencer accounts, but slicker. Cannon explains the rationale: 'You follow a person's social media and you want to emulate their life.' Which means buying a bottle of Folc. It's not particularly original, but it's effective. And it's clearly working. Folc is just a small brand, producing only 15,000 bottles in 2022, but it has plans to double its output every year.

The other 'pure brand' (without vineyards or winery) in the English wine industry is The Uncommon. It was founded by Alex Thraves, a graphic designer, and Henry Connell, a former finance man turned winemaker. Rather than plant a load of vineyards in Sussex, they set up a virtual brand and sold wines sourced from around the country. The first release was a bacchus made at Denbies in Surrey. Everything about the brand was different. Rather than a bottle, the wines come in cans, tapping into the burgeoning market for less energy-intensive packaging (glass takes a lot of energy to make and transport). Then there's the branding, with the first release featuring a dapper giraffe sporting a bowler hat. It's been joined by a sparkling rosé and a spritzer. The look is witty and distinctive, and it helps that the wines taste good as well.

It's often the smaller producers who have the strongest image. Cannon rates Tillingham highly, with its 'East London comes to Sussex' vibe. A photographer friend also praised Tillingham, saying that the branding, the social media, the bottles and the visitor experience were all spot on (though he wasn't quite so keen on the wines). Or there's Westwell in Kent. The winemaker, Adrian Pike, used to be in the music industry and his artist wife Galia designs the distinctive labels. Each one is a little artwork – like an album cover. There's a music business link with Gutter & Stars in Cambridge too. Winemaker Chris Wilson names the wines after songs – I Wanna be Adored and Blonde on Blonde – perfect for the ageing hipster market. Everything from the wines to the labels to Wilson's social media presence seems entirely natural. It doesn't feel like it's been through focus groups or a marketing agency, because it hasn't. But it isn't haphazard either. The wines are even packaged with branded Gutter & Stars packing tape. Not all smaller producers do the branding so well. Wine shop Bin Two in Padstow produces pet nat[78] wines called 'Fizzy Bum Bum.' No thank you.

At the moment, most English producers are concentrating on the strong domestic market, but what about export? The figures are currently tiny for the industry as a whole, with around 92% of English wine sales in the UK. Gusbourne exports about 15 per cent of its production but Charlie Holland wants to increase this to hedge against a vulnerable UK hospitality industry. According to Simon Roberts at Ridgeview in Sussex though, export requires a lot of work, with regular visits to major overseas clients. Ridgeview sells to Scandinavia and the far east, in particular Japan and Taiwan. This sales model was badly disrupted by the pandemic.

For Harrow & Hope in Buckinghamshire, as with many English wine producers, the Scandinavian market is the most reliable. 'Scandinavia has a taste for English wines,' Henry Laithwaite says. Norway, Sweden and Finland all have alcohol

---

78 See 'Fizz Wars', chapter 8

monopolies, meaning English producers submit their wines and if they meet with approval, they are sold across all state-owned shops. Then the main way for English winemakers to promote their wines is through state-owned wine magazines. 'It's an odd way to sell wine,' Laithwaite admits, but if you can break in, it's very straightforward.'

In contrast, the American market is incredibly complicated, with each state having its own laws on selling alcohol. Even some Californian producers find it's not worth selling their wines in certain states. Furthermore, the distance and work required to sell English wines in export markets mean that English Sparkling Wine is competing head on with Champagne. Laithwaite says that sales have still not returned to pre-Covid levels but hopes that the recent drop in value of the pound might make English wines look like good value. Every silver lining and all that.

One brand which has been surprisingly slow off the mark with the export market is the biggest of them all – Nyetimber. Eric Heerema explains: 'We focused on our international market later than some of our English sparkling colleagues. We wanted to focus on building a brand in this country first. There's not much space in international markets. In each market that we enter, we need to convert a certain number of people from Champagne to Nyetimber. That's a very gradual process. But in the last two years, exports of our wines have risen enormously and look very good, even within Europe, in spite of Brexit.'

Mark Driver at Rathfinny agrees that 'it's hard work selling wine,' but believes that Britain has a strong image and the British are 'well-liked, at least outside Europe'. British sells, he says. The Simpsons also believe that foreign markets are receptive to English wine. They told me of a Dutch wine merchant, 'the

Berry Brothers of the Netherlands', which refused to stock the couple's Languedoc wine, Domaine Sainte Rose, but 'wrote to us and said [we] want to be your importer' after the head of the company heard about the reputation of English wine.

Christian Seely from Coates & Seely's main role is as CEO of AXA Millésimes, which owns such reputed estates as Quinta do Noval in Portugal and Pichon-Longueville Baron in Bordeaux. It takes him all over the world, and he finds that wherever he goes, people are curious about English wine and delighted to try it. Coates & Seely currently exports about 25 per cent of its production; Seely was particularly pleased to get his wine into Michelin-starred restaurants in France including the George V, Le Bristol and Alain Ducasse's Spoon in Paris. Now there's something you couldn't have predicted 20 years ago – the French buying English wine.

CHAPTER 17

Storm clouds ahead

'The reality is that the pandemic has fractured production lines. It has re-regionalised, sometimes re-nationalised, certain production chains. And I think it has permanently de-globalised a large part of world production.'

Emmanuel Macron

I n the spring and summer of 2022, Ukrainian flags were everywhere, flying from pubs, garages, private houses and town halls across England. In Faversham, the Kent town where I live, there were far more Ukrainian flags than Union Jacks or St George's Crosses. A sign of solidarity with the Ukrainian people but also a reminder that with a war raging in Europe, a lot of things we have taken for granted for a long time are now being called into question.

One of the world's largest bottle manufacturers, Verallia, has a factory at Zorya in western Ukraine which shut down when the Russians invaded. The problem is, of course, trivial compared with the suffering of the Ukrainian people but the knock-on effect was a desperate shortage of glass in England. Charles Simpson of Simpsons in Kent explained that it takes weeks for glass-making kilns to warm up, so even when new factories are up and running, most bottle producers are concentrating on the more profitable heavy bottles. 'It's a nightmare,' he says. And it's not just Ukrainian glass that has been affected by the war; glassmakers globally have been hit with soaring energy prices too. Felicity Carter wrote in *Wine Business* magazine: 'Of all the industries hurt by the loss of cheap Russian gas, glass has been one of the most affected. The Continent hosts 162 glass plants across 23 countries, all of which must run continuously, and most of which are powered by gas.' She went on to say that larger industries like car makers and breweries have been stockpiling glass, further driving up prices. Every producer I visited in the spring of 2022 expressed concern about getting hold of bottles; and if they can't bottle the 2021 vintage then they won't have space to store the 2022 when it comes in. At Chapel Down, head winemaker Josh Donaghay-Spire had to break off from my visit to field a fraught phone call with a supplier.

It's not just glass, either. Everything seems to be in short supply. Most of the paper used to make bottle labels comes from a factory in Finland where there is a long-running industrial dispute. A strike in Finland means a delay on receiving the labels in England. The Covid lockdowns haven't helped. Who could have guessed that after shutting it down, turning on the world's economy again wouldn't be straightforward? As no less a figure than French president Emmanuel Macron said: 'The reality is that the pandemic has fractured production lines. It has re-regionalised, sometimes re-nationalised, certain production chains. And I think it has permanently de-globalised a large part of world production.'

And that's before you get to the 'B' word. Most English producers are not happy at the complications that Brexit has caused. Simon Roberts from Ridgeview is quietly furious: 'There's nothing good about Brexit,' he says. The problem is that almost all the technology needed to make wine comes from France, Germany or Italy. Roberts told me about some machinery parts that he had ordered from France which then sat at Paris' Charles de Gaulle airport for four weeks while customs waited to clear them. Mark Driver from Rathfinny is more sanguine. He makes sure he orders things between six and nine months before he needs them and so hasn't had much of a problem – yet. But the 'just in time' approach has disappeared, he admits. The Simpsons are in a fortunate position of having a winery in France so they can order things directly from across the Continent and have those parts sent to their winery there. Charles Simpson explains: 'Post-Brexit, cork and glass suppliers won't send to England but they will send to France.' Every month or so, the Simpsons put together a container load and have it sent over. Nevertheless, 'costs of production have gone through the roof,' he adds, estimating that it costs around an extra £100 a pallet[79] which has to be added to each bottle.

All this means extra headaches for producers – and then there's the worry about who is going to pick the grapes. In the past, a steady supply of Romanian and Bulgarian workers would be available to do this. Both Romania and Bulgaria are

---

79 It depends how big the pallet is, how thick the bottle glass is and how the bottles are packed, but on average a pallet load will be about 1,000 bottles

grape-growing countries and skilled, efficient pickers were relatively easy to come by until 2019, when the United Kingdom
finally left the European Union. Suddenly things became
a lot more complicated. Some estates like Harrow & Hope in
Buckinghamshire are lucky in having a nearby group of
Romanians with residency status. For others, it has been harder.
Simon Roberts from Ridgeview described 2021 as 'an absolute
nightmare. There were no pickers.'

It's generally the smaller estates that find it easier as they're
able to marshal local people who will often do the work because
they enjoy a day or two out in the vineyards. Picking at Breaky
Bottom is a social occasion and Peter Hall's wife Christine lays
on a good lunch with lots of wine as an extra incentive. Clare
Holden from Brissenden also taps into the community for all
her picking needs, recruiting parents from her children's school
and the local football club. She sees 'the same old faces' every
year – and not just for picking; a team of local mothers helps
out with all kinds of vineyard jobs. 'We all like the therapy of
being outside,' says Holden. 'It's nice to go along the vines with
like-minded people. The jobs can be pretty mundane and
repetitive but you need skill and knowledge.'

Other vineyards are being even more creative when it
comes to finding workers. Yotes Court in Kent advertised grape
picking as a 'vineyard experience' at £35 a pop. Yes, you pay
to pick their grapes. For that you get, 'a morning in the vines
with lots of grape-picking' but also breakfast, lunch and the
promise of a bottle of the wine 'made from the grapes you
picked when it is released'. Admit it, you're tempted. Meanwhile at Langham Estate in Dorset, Fiona Wright, who looks
after sales and marketing, had the brilliant idea of combining
grape picking with dating. I'm not sure she was being entirely

serious, but what a great way to get people into the vineyard.

Sadly, there are only so many volunteers or people who will pay for the privilege of picking grapes, so larger estates have had to come up with something new. For Mark Driver at Rathfinny, that has meant using local workers. Winemakers might have to pay them more and they might work more slowly, but there are people in the English countryside who are able to do some part-time work. It's a similar story at Gusbourne, where Charlie Holland told me, 'We've gone back to using local people. We have people here who have picked since day one, like hop pickers in the old days. Vines are the new hops.' Mark Gaskain at Domaine Evremond is looking further afield. Last year they used Ukrainians, Belarusians and Russians but for obvious reasons that's not possible in 2022, so he was planning to use labour from Nepal and Indonesia, whose workers he described as 'very agile'. He explained: 'We have technical full employment in the South East so it's hard to get people to work for four weeks. We need these people. We'd be stuffed without them.' There's a company called AG Recruitment based in Faversham which has been recruiting workers from Asia as part of the Seasonal Agricultural Worker visa scheme – which explains the changing faces of the itinerant farm labourers waiting to be picked up from Tesco car park over the summer.

It seems crazy that workers from thousands of miles away can obtain working visas but those from the European Union cannot. There are also stories circulating about Indonesian workers being exploited. An investigation by the *Guardian* alleged that some workers were being charged exorbitant deposits for their working visas by agencies in Jakarta, leading to many being in a form of 'debt bondage'.

As with many things that are laid at Brexit's door, leaving

the European Union exacerbates an underlying problem – the Europe-wide shortage of cheap, semi-skilled labour of the sort needed to pick grapes. I visited the Mosel Valley in Germany in 2017 and growers there were moaning that they couldn't get the Romanian labour like they used to. One nearly caused a diplomatic incident when he complained that the Polish pickers he had hired ended up drinking too much – not realising that two of the journalists in the room were also Polish. That year, to make up for the missing Romanians, the Mosel Valley growers were using a mixture of Syrian refugees, students and wine writers to pick their grapes. When I saw that grape picking was on the schedule for my trip, I was expecting to be shown a few vines, maybe fill half a bucket with grapes, before heading inside to taste some wines. Instead, we were up at six in the morning, handed some secateurs and driven to a damp, steep vineyard where we picked grapes until lunchtime. If they're entrusting journalists with such work, these people must be desperate.

We weren't just there to help with the harvest, though. We were also given a demonstration of the latest generation of picking machines which it is hoped will solve the problem. These tractor-like machines were terrifying examples of Teutonic efficiency that thundered down the steep slopes of the Mosel Valley with an uncannily steady tread. They looked like something out of *Terminator 2: Judgement Day*. Picking machines aren't new, however. They've been around since the 1960s and are useful in places like much of Australia or the coastal plains of the Languedoc that are relatively flat, have lots of vineyards and a shortage of labour. They work by shaking the vines so that the grapes fall off. *The Oxford Companion to Wine* describes the most common kind of picking machine as a 'horizontal slapper', which always makes me laugh.

The problem that picking machines have is that they tend to be quite crude, grapes get smashed up and by the time the fruit has reached the winery – which might be days in a country as big as Australia – the juice is beginning to oxidise. You also get what's known in the business as MOG (material other than grapes), which could be anything from bits of wood and leaves to birds and small mammals. Not what you want in your prestige cuvée. So there's something of a prejudice that high-quality wines cannot be made from machine-picked grapes. Furthermore, to make sparkling wines via the Champagne method you usually need whole bunches, where the grapes are more likely to be intact. This helps stop the juice taking on colour and other undesirable compounds from the grape skins. But machines cannot harvest whole bunches so for the timebeing, to make England's speciality, sparkling wine, you still need human pickers.

Nevertheless, the machines are coming. Sam Barnes of Biddenden has a business that supplies vineyard services and equipment, including picking machines. He has two Pellenc 890 machines. When I asked him how much they cost, he replied: 'too fucking much'. That translates as around £130,000 each. You also need a tractor to pull them, ideally one with an automatic transmission so you can minutely control your speed. That's another £70,000, although you do at least then have a tractor that you can use for the rest of the year. And machines have some advantages over people. They're cheaper (once you've paid off the initial investment) and they're much faster. This is especially important in a marginal climate like we have in England as it means that growers can take greater risks with grape ripening, knowing that if the weather turns, they can harvest quickly.

Barnes didn't give me the sales pitch, though. 'There are pros and cons,' he said. 'Machine harvesting depends on the wines you are trying to make, the price point and the winemaker.' He charges by the hectare, so if your yields are very low, with not many grapes per vine, then you're probably better off with human pickers. But as yields go up, the machines start to make more sense. Barnes estimated that, with an average yield, a producer's cost will be around £400 per hectare using pickers and £80–90 using a machine. Plus the machine can then be driven with the fruit directly to the winery.

In a very bad year like 2021, when many growers suffered from rot, picking by hand would be prohibitively expensive as pickers would have to cut out the damaged berries. With a machine picker, in theory only the healthy berries would be shaken off. But certain grape varieties, notably those with thicker skins, are able to handle the machine picking process better, whereas more delicate grapes might get broken up and arrive at the winery wet and potentially oxidised – though that's not such a problem in England's cool climate as it would be in southern France or Australia. It is, Barnes says, possible to make sparkling wine from machine-picked grapes but he admits it's not ideal. 'That's at the winery's discretion. You've got to have the right winemaker. It's a slightly dirtier job.' At Chapel Down the winemaker told me that while he is happy to use machine-picked fruit for still wines, he isn't for sparkling.

The environmental impact of machines versus people is hard to quantify. The machine is, of course, powered by diesel, which is not great for your carbon footprint but then again neither is flying workers over from Nepal. So as with most of these things, there are trade-offs.

The breakdown of globalisation doesn't just mean the end of

cheap labour and easy access to quality French equipment. Part of the reason for the vast improvements in English wine are to do with home-grown winemakers previously working in countries like New Zealand, Australia and South Africa, and overseas winemakers coming to England. The pandemic-induced disruption to reliable, affordable air travel put a spanner in the works. New Zealand only opened up again in 2021 after abandoning its zero Covid policy. And of course Brexit makes things complicated for European winemakers wanting to work here and vice versa.

Not everyone, however, is so pessimistic about the effects of Brexit. Mark Driver of Rathfinny caused consternation in the small world of English wine by saying that there might be upsides to not being in the EU. 'Brexit helps us. It offers opportunities to take a different path,' he argued. One such opportunity might be offering his sparkling wine in a pint bottle as Champagne producers like Pol Roger used to do in the past. Apparently this was Winston Churchill's favourite format. 'Perfect for two people at home,' Driver said. It seems a mild enough suggestion, but he described the reaction from fellow winemakers and especially journalists as 'toxic'.

But he's not the only one looking on the bright side of Brexit. When he was chairman of industry body Wine GB, Simon Robinson from Hattingley Valley kept a scrupulous neutrality during the rancorous Brexit arguments that played out within the trade. He then claimed, in an interview with industry title *The Buyer* in 2019: 'Brexit and a weak pound will probably help significantly,' in terms of shifting the wines that were coming on to the market. His sales and export manager Gareth Maxwell agrees: 'Brexit and global warming have been good for us, I'm afraid,' he said in an interview with the *FT*. Henry

Laithwaite of Harrow & Hope reports that the pound's woes in autumn 2022 led to more enquiries from abroad. But of course a weak pound means the equipment needed to keep the industry going becomes more expensive.

The flipside to problems with globalisation is an increased interest in localism. Many middle-class shoppers, who make up the bulk of the market for English wines, are keen to cut back their food miles. Whether this is actually more environmentally friendly or not hardly seems to matter; even though it might in some cases be more energy-efficient to grow something in warmer climates and ship it to Britain, it doesn't feel like the right thing to do. Furthermore, people like to feel a connection to where they live. Pierre Mansour, head of wine at the Wine Society, one of Britain's biggest mail-order wine merchants, said that customer feedback suggested people are prepared to buy local products even if they cost a little more. Richard Chamberlain, who runs the Craft Drinks Company, has built a business on supplying Cotswolds produce like beer, cider and even the very fine local whisky. Part of what drove the gin boom is that people like the idea of having a Yorkshire gin or a Manchester gin. This regionality is something that English wine has only just begun to tap into.

For Stefano Cuomo, whose family own food hall Macknade in Faversham, local is more complex than just proximity to producers. Customers have to be invested in the product and the producer, he says, citing the example of Tom Berry of Brenley, a local wine producer. The Berry family had been growing hops in Kent for years, before recently switching to grapes. They have just begun producing wine under their own label, made at Defined in Canterbury. Cuomo's customers can meet Tom, who comes into the shop in his muddy wellies rather than the branded gilet that is synonymous with corporate winemakers. The owner is not some big faceless enterprise run by City boys, but a family who drink in local pubs. When it's like this, and the wines are good, then people will feel invested in local producers, says Cuomo.

But are there going to be enough people prepared to pay extra to buy their wines? That's a harder one to predict. With rising inflation and wages that aren't keeping up, most British people are poorer than they were five years ago. Is the trend for localism a fad for the wealthy while everyone else goes back to buying Aldi's excellent vinho verde at £7 rather than the English equivalent at £17? The average British wine drinker is notoriously cheap. Tim Phillips from Charlie Herring wines thinks that we have our priorities all wrong. He points out how many people live in houses worth £1 million but won't spend more than £8.99 on a bottle of wine; in the US people will often spend $50 on a bottle for the weekend. Henry Laithwaite, however, is bullish: 'My dad has been through some recessions, and alcohol is one of the last things that Brits give up.' From my own experience, he might be right. During the financial crash of 2008, an upmarket wine merchant opened up near my then home in Hackney. There was very little on sale below £10. I remember giving it six months before it would go out of business. Instead, it thrived and the manager explained to me why. In a recession, people stick with the everyday luxuries but they cut back on the more expensive stuff. So rather than going out for dinner in Soho, local customers were getting a takeaway and a nice bottle of South African chardonnay.

The challenge among producers is to persuade more and more customers to make the switch to English wine. Mark Driver observes that in the UK, we 'consume five million bottles of wine per day, but produce only five million per year'. The fizz market alone is huge: in 2021, 29 million bottles of Champagne[80] were imported into Britain. Much of this, however, will be very price-sensitive supermarket own-labels, something with which English fizz can't compete. The truth is that the British market for wine is in decline, with only rosé and Prosecco seeing growth. Yet the planting in England continues. Vines take about three years to produce a proper crop and then English Sparkling Wine needs at least eighteen months in bottle, so it's likely that in five years' time an ocean of English fizz will hit a shrinking market. There is talk about an impending English wine glut, just as happened in Australia and New Zealand in the nineties and noughties. It might already be happening. Mike Best from Boutinot wine merchant claims that there are 'producers with a lot of 2014 in the cellar without a plan or a home for it'. This is something you notice at wine tastings, where the vintage being poured is sometimes an old one, suggesting that there might be problems getting through stock.

We are likely to see more wines like Morrisons' own-label English Sparkling Wine, the 2010 vintage of which came on to the market in 2022 for £25 a bottle. The rumours were that it was from Nyetimber, and either the result of an overstock or that the vintage was somehow not up to scratch, so the producer didn't want to market it under its own label. Whoever made it, the price was very low for a 12-year-old English Sparkling Wine. It won't just be one-offs like this, though; oversupply should

---

80 Champagne governing body the CIVC (Comité Interprofessionnel du vin de Champagne) rather shot itself in the foot in 2020 when, because of falling demand during Covid, it restricted yields, the amount of grapes allowed per hectare, so as to avoid a potential surfeit and protect prices. But as the world recovered from the pandemic, demand soared which Champagne producers won't have the stocks to fill. This is all great news for English growers

mean a price drop for most wines. Not great news for producers, but good for customers. Charlie Holland from Gusbourne thinks 'the supply of grapes may get too much and grape prices will therefore go down'. Such a trend will be of benefit to urban wineries like Gutter & Stars which depend on buying in fruit.

It's not just customers that may be in short supply in the future. What about capital? As we saw in the Money Men chapter, the boom in English wine has been partly driven by low interest rates, something that has changed in the last year. What happens when the cheap money dries up? Most English wineries are not yet profitable. The owners have invested in the expectation of turning a profit in the long term. But with the supply of money becoming expensive, some investors might decide to cut their losses. Looking through the accounts of wineries on Companies House makes fascinating and worrying reading. Many are technically insolvent, losing money, just a step away from a creditor or bank calling in its investment and going bust as a result. As Stephen Skelton puts it: 'There will be some casualties, especially those low-cropping, poorly performing sites which are only held together with the glue of money that their current owners are providing.' Right now, we're seeing a rash of closures in the gin and craft beer world, both of which enjoyed huge periods of growth in the last two decades. You would be brave to bet against a similar thing happening to some English wineries.

towards Canterbury, and the weather was very different, hovering around freezing with a dark, ominous sky. After interviewing Ruth and Charles Simpson, I went out into the vines with Henry Rymill, a cheerful Australian who looks after events at the property. We'd only been out for a couple of minutes when the snow came in horizontally and he had to shout to be heard. It was, in any case, too cold to write and my notebook was getting soggy so after 15 minutes he suggested we go in and taste some wine. I didn't protest.

Warmer winters and cold springs are becoming increasingly common as the climate changes. It's a combination that can be devastating for vines. A warm winter encourages the vines to come to life earlier but this makes the delicate buds susceptible to a late spring frost. The effects are quite startling. After a severe frost, there are blackened, lifeless patches on the vine where grapes would otherwise have grown. Chardonnay is particularly susceptible as it buds early. Henry Laithwaite at Harrow & Hope in Buckinghamshire lost 70 per cent of his crop this way in 2016. The following year, 2017, was 'the worst ever for frost', he said, but this time he was prepared. Lighting large candles (known as 'bougies', the French word that to my ear sounds like a nightclub in Clapham) in the vineyard as the temperature drops is effective at protecting the vines from frost. It's expensive though. Bougies cost around £8 each and you need hundreds for even a small vineyard (plus staff to position and light them). Not only are they expensive, but as French products, their supply to England has been disrupted by Brexit, the war in Ukraine and Covid. At Simpsons they also have huge blowers which move the air around and help mitigate against frost (in California they use helicopters), but according to Laithwaite they are not as effective as the old-fashioned candles.

I've always thought frost was just cold weather but it's more sinister than that. Laithwaite describes it as a 'gloopy liquid' that rolls down hills and if it meets something to stop it, it 'piles up twice as high as the barrier', which made me think of something from a James Herbert novel or a fifties sci-fi film, *The Frost!* Ditches, hedges and low-lying land can all provide frost traps in the vineyard. Thankfully, in 2022 it had warmed up again by the time of budburst in April and most English vineyards were not severely affected. The French weren't so lucky, with growers in the Rhône and Languedoc particularly badly hit. If you're like me and follow lots of wine producers on social media, in April your feed will often be full of images of candles burning at night in vineyards. Beautiful to look at but terrifying for growers.

Laithwaite describes frost as 'part and parcel of global warming', in that warmer springs and earlier ripening mean there's always the risk of frost. Despite winning awards for its wines, one vineyard in Oxfordshire, Bothy, decided to give up completely in 2020 as it was losing so much production to frost that it wasn't profitable to keep going. Inland vineyards like Bothy and Harrow & Hope are hit worse than ones near a large body of water, as the water stops the temperature from dropping too low. Laithwaite thinks it's worth the risk though, as the warmer summers of inland locations mean riper grapes, but he admits that successful winemaking is only really possible with a small vineyard, where producers are able to stay on top of things with bougies. Tommy Grimshaw at Langham in Dorset, in contrast, was dismissive of the issue, saying those inland producers having to mitigate against frost were simply located in the wrong place.

Articles on the growth of the English wine industry will

often begin by talking about climate change. I've decided to end with it because I wanted this book to be about people first and foremost. I'm largely in agreement with Kent and Sussex winemaker Will Davenport who told me that the 'change from amateurishness is more important than the changing climate'. Wine journalist Tom Hewson agrees, saying: 'Climate change alone doesn't make better wine; it only gives England the opportunity to do so.' Yet in his 2022 report on English wine, Hewson points out that the levels of ripeness seen in recent vintages would be 'unimaginable even in the eighties and nineties'.

There's no denying that, frost risk aside, the changing climate has been largely beneficial to English wine. In 2014 the Conservative politician Owen Paterson upset many when he said just this about English agriculture, though most growers in England quietly agreed. Pierre-Emmanuel Taittinger added that: 'We now have many very good English sparkling wines. Global warming helps, as the climate is more generous.'[81] And while Gusbourne's Charlie Holland admits that 'being a poster boy for climate change is not a good look', he is fully aware of how global warming has enabled him to make better still wines. In the first decade of the century, Gusbourne had three good years for still wines; in the 2010s it had four. 'That's the direction of travel,' Holland said. Bob Lindo of Cornwall's Camel Valley cheerily adds: 'Climate change is obviously a creeping catastrophe, but in the short term, it opens the door to more grape varieties and higher yields.'

The figures speak for themselves: the industry has expanded from a negligible number of vineyards 30 years ago to over

---

81 From an interview in the *Daily Telegraph*

4,000 hectares of vines today. There's been a shift from producing small quantities of wines made from Germanic grapes to everything from Champagne-style sparkling wine to rosé to deep-coloured reds.

The expert on climate and English wine is Dr Alistair Nesbitt who did his PhD in viticulture and climate change at the University of East Anglia (UEA) and now runs Vinescapes, a vineyard and winery consultancy business. He forecasts: 'In some areas of the UK, the bumper vintage of 2018 will become the norm, and the Champagne region's grape-growing temperatures from 1999–2018 are projected to occur across an expanding area of England during 2021–2040. In certain years, a few areas of the UK may see growing-season climates similar to those that contributed to the very best recent vintages of Champagne, as well as supporting increased potential for Burgundy and Baden[82] style still red wines.' So present-day Champagne temperatures in the near future and more decent reds . . . Sounds promising, no?

At the moment, the temperature in Southern England is around what it was in Champagne in the seventies, but the two can't really be compared as England's maritime climate is much wetter than Champagne's continental (inland) climate. As John Atkinson from Danbury Ridge in Essex explains: 'Light levels don't track temperature, and England will remain a maritime climate, with a lag in peak temperatures relative to the continental distribution experienced by much of Europe, where June qualifies as a summer month.' In other words, summers start later in England and will continue to do so. This damper climate means that yields are always going to be lower than on

---

82 Area of Germany that makes increasingly good red wine from pinot noir

the Continent and will continue to vary dramatically between vintages. Even warm years like 2022 are not without their problems. When I visited Danbury Ridge, Duncan McNeill, Mr Essex himself, was worried about the lack of rainfall from June to September. 'We were down 10–15 per cent in rainfall, which is worrying. It's drier than Burgundy.' Of course it then poured with rain for most of the autumn – another sign of how the climate is changing, and a harbinger of its own problems, not least fungus at harvest time.

All over England, growers were worried about the lack of rainfall during 2022. Older vines are more resilient as they have deeper roots that can find water but younger vines need watering. According to Wine GB,[83] some English vineyards have now fitted drip irrigation systems which carefully regulate the amount of water the vines get – something I never thought I'd see in England. Tommy Grimshaw at Langham thinks that irrigation has a helpful side effect, in that it can be used to combat frost: 'Irrigation has some interesting anti-frost properties in the early spring,' he says.

The changing climate has seen English winemaking move away from the Germanic varieties that were the mainstay of the industry in favour of their French counterparts. As we saw in the Eastern Promise chapter, Danbury Ridge makes startlingly ripe still wines from chardonnay and pinot noir. With a warming climate, these might become the norm in southern England. As Charlie Holland of Gusbourne says: 'We may want to move away from sparkling to still. Not in 5–10 years but further in the future. We want to be making the best wine we can here, and these are options.' We are certainly likely to get more still wines,

---

83 *Drinks Retailing News*

which need riper grapes. Another producer that is expanding its range in that direction is Balfour in Kent. Winemaker Fergus Elias says he finds still wines much more exciting. 'Sparkling will likely remain the "hero" brand as Wine GB likes to refer to it but I expect a lot of producers will be pushing out increasing quantities of still wine.'

Although Simon Roberts thinks that Ridgeview could have made still wines in 2018 and 2020, he hasn't taken the plunge yet. 'Ask me again in ten years' time,' he said when I asked him whether he had any plans to do so. But another sparkling specialist, Hattingley Valley in Hampshire, did produce some still wines in 2020, so some sparkling wine producers are making the switch. I'd be very surprised if most, if not all, sparkling specialists weren't making some still wines in the next five years, even if just to sell at their cellar door or restaurant.

The big question, however, is whether producers can make *high-quality* still wines every year. According to Duncan McNeill, who has been working in vines in East Anglia since 2006, the emblematic year for still wines was 2021. It was a damp, overcast summer, yet a warm autumn meant that he was able to harvest quality pinot noir from Danbury Ridge at 13% ABV. If they could make a good red in 2021, he thinks they can probably do it every year. This wasn't just the case in Essex; I tasted some fine pinot noirs from around the South East from the difficult 2021 vintage.

All in all, the English climate is looking very positive for winemaking right now. But how might it look in twenty or even forty years' time? Vineyard and winery consultant Alistair Nesbitt has produced twelve climate models to try to predict how the climate is going to change in the near future. His research is based on something called RCP[84] 8.5 produced by the IPCC, which Nesbitt calls the 'worst-case scenario' for how warm temperatures are going to get. There are other RCPs which aren't quite so dramatic but, according to Nesbitt, this RCP is 'currently tracking globally'. Based on this, he thinks global temperatures will get between 1–1.2°C warmer by 2040. He forecasts, however, there will be 'no significant increase or decrease in rainfall. It really is all about temperature.'

Looking even further ahead, the wine merchant Laithwaites commissioned a study by Professor Mark Maslin and Lucien Georgeson from University College London to look at how climate change might change things by 2100. It suggested that we might be growing warm-climate grapes like syrah and tempranillo in London, while in the Scottish borders riesling and pinot gris could be thriving. It all sounds rather fanciful, but there are already commercial vineyards in Yorkshire, Derbyshire and even one in Scotland. Could a Côte de Trossachs be a wine of the future?

Climate change does mean that long-standing viticultural regions are having to think about using different grape varieties. Nebbiolo in Barolo and Barbaresco is regularly nudging 15% ABV while some parts of Burgundy may at some point become too hot for pinot noir. In Australia, Italian, Greek and Spanish varieties are becoming increasingly popular as they are better able to handle the heat, while in 2021 the Bordeaux authorities allowed the planting of foreign varieties including touriga nacional and alvarinho from Portugal. While the changing of varieties might be a marketing disaster in long-established regions, since people associate Burgundy with pinot noir and

---

84 Representative Concentration Pathway

Bordeaux with merlot and cabernet, it's an opportunity for a country like England, about which very few people have preconceived notions.

Fergus Elias is relishing getting to grips with more varieties. 'The varietal mix will change, with less seyval – God willing – and less Germanic varieties as a whole,' he says. 'I feel the Geisenheim[85] model will be supplanted by something distinctly more Burgundian in shape and style. Lots of pinot noir and chardonnay, but also varieties such as gamay, albariño and a little riesling, plus maybe some PiWis if any of them actually turn out to be nice.' The perennial problem with PiWis, Elias says, is their names. 'Very few consumers will choose a souvignier gris over a sauvignon blanc or pinot grigio.'

English wine expert Stephen Skelton, however, doesn't think the varieties will change that much any time soon: 'Viticulture moves forwards pretty slowly. If you have vines in the ground producing good wine, why would you replant?' He thinks the industry's short-term future is going to be all about making still wines from chardonnay and pinot noir. 'If you can grow, make and sell still wine from these varieties at £20–£35 a bottle (the same as commercial Chablis, Burgundy and New World examples) and make good profits, then this is what many people will do.' While some growers are experimenting with riesling, sauvignon blanc, albariño and even merlot, shiraz, pinotage and cabernet franc, 'It's a real risk to plant too much of speculative varieties when you can plant standard established varieties and more or less guarantee good wines.'

Nevertheless, growers in England are experimenting with warmer-climate varieties. There's a little albariño in Kent from

85 German wine school, see 'The Bloody Awful Weather Years', chapter 2

which in good years Chapel Down makes a great wine. The winemaker, Josh Donaghay-Spire, recounts with pride how in a blind tasting growers from Galicia in northern Spain thought it was a local wine. Even more ambitiously, Bolney in Sussex planted some merlot, though without much success. At the Wine GB tasting in September 2021, no fewer than three growers told me that they had the only sauvignon blanc in the room. All three weren't bad.

Don't write off the old varieties just yet though. However the climate changes, Skelton thinks that 'our low yields and therefore low profitability is a real issue and is one of the big things holding expansion back'. His money is on a renewed interest in less glamorous varieties with higher yields 'such as seyval blanc and reichensteiner, plus some of the PiWis'. It's also not just about what you plant, but where you plant it. 'At Danbury Ridge,' says Atkinson, 'we've invested a lot of time in finding areas that are buffered from the vicissitudes of weather – seasonal and long term – because, as Mark Twain put it, "Climate is what you expect, weather is what you get".'

Apart from planting suitable varieties, how can growers temper climate change? Peter Hall at Sussex's Breaky Bottom, ever the maverick, doesn't think that there's much we can do. He thinks we're 'not giving credit to natural climate change. There are warmer and colder times. We are in an interglacial period. It's getting warmer anyway.'

Most growers are united in their concern for the amount of carbon they put into the atmosphere. Various wineries, such as Hundred Hills in Oxfordshire, describe themselves as carbon neutral. To burnish its environmental credentials, Ridgeview in Sussex has gained B Corp accreditation. This comes from a non-profit company called B Labs that marks businesses on their 'benefit for all', hence the 'B' descriptor. Sarah Driver from Rathfinny, which was B Corp-accredited in April 2023, says: 'B Corp is regarded as the gold standard when it comes to sustainability and it believes in using business as a force for good.' The scheme is not just about environmental impact. Companies are also marked on their community engagement and diversity. But B Corp accreditation is not without its critics. Some have described it as akin to 'greenwashing' – making businesses appear environmentally friendly purely for PR purposes. B Corp doesn't publish the results of its impact assessments, so it's not clear exactly what its criteria are for companies to gain accreditation. Its certification is also not legally binding, so if a client doesn't do what it's supposed to, the only thing B Corp can do is to remove the company's B Corp status, like it did with maverick brewer Brewdog in 2022.

The rigour of B Corp certification is somewhat undermined by the fact that Nespresso became certified in 2022. This is despite controversy over child labour, wage theft, deforestation and the fact that Nespresso is based on a single-use aluminium pod which doesn't sound particularly sustainable. Other notable B Corp companies include Innocent smoothies and Ben & Jerry ice cream, both of which are owned by multinationals. There's something rather 'Davos man' about the idea that giant corporations are the answer to the world's problems. Some, like Anand Giridharadas, author of *Winners Take All: The Elite Charade of Changing the World*, see it as something of a club for rich companies. He writes: 'A new breed of community-minded, so-called B Corporations has been born, reflecting a faith that more enlightened corporate self-interest – rather than, say, public regulation – is the surest guarantor of public welfare.' B Labs charges anywhere between $500 and $50,000 for accreditation depending on the size of the company it is auditing.

B Corp accreditation is enormously useful, though, as an answer to what companies are doing about the problem. That's not to say that Ridgeview or Rathfinny are anything less than totally sincere and committed to being more sustainable. In a time of increasing energy prices, worries about carbon emissions and potentially a shortage of water, being as efficient and economical as possible is clearly a good idea. As is paying your staff properly and including them in the business. But it's worth remembering that just because a company has B Corp status doesn't necessarily mean that it's actually better for the environment than a non-B Corp company.

Whether any of this really makes a difference to global warming, however, isn't entirely clear. Adrian Pike from Westwell in Kent says: 'It's really hard to know when governments all around the world don't seem prepared or in a position to attempt to slow things down.' Charles Palmer, owner of a sparkling wine producer in Sussex, comments: 'Don't think for one moment that anything that wine producers do will actually slow down or even affect climate change. The main drivers of CO_2 emissions are population growth, deforestation, massive industrialisation, and global commerce. The part that viticulturists can play is caring more for the local environment, not pretending that it's possible to turn the tide of climate change.'[86] I can't help thinking that even if the entire English wine industry went carbon neutral, it wouldn't make the blindest bit of difference when China and India are building dozens of coal-fired power stations every year. But perhaps I'm being cynical. Alistair Nesbitt disagrees forcefully: 'It's not pointless, absolutely

---

86 From LinkedIn

the opposite. If we don't do something, globally the temperature will keep rising to a point where viticulture becomes unviable in parts of the word.'

Cutting down on packaging is one way of lowering emissions, as Tommy Grimshaw from Langham explains. 'Most vineyards are carbon negative due to the vines, but the big carbon impact comes from the winery. About 60 per cent of this is from the glass,' he says. Whereas other packaging such as cartons, aluminium cans and kegs do exist, to make English Sparkling Wine you need glass bottles. These need to be fairly heavy, though growers such as Will Davenport have moved to lighter bottles without any problem. So let's all recycle! Grimshaw is sceptical about this too: 'Recycling glass isn't the answer as it takes almost as much energy to melt the old bottle down as it does to make a new one from scratch.' Instead, he argues, we need to look heavily at re-using glass. There are a number of hurdles in the way with this too, though. 'Firstly, the only machine that can remove labels and sterilise the inside of the bottles costs hundreds of thousands of pounds and is absolutely huge. Secondly, there hasn't been enough research into the safety side of reusing glass. Bottles get knocked around quite a bit in transit and in people's homes, so there could be a bottle weakness that would need to be taken into account.' There are now bottles made of paper. Laithwaites produced a bacchus that came in a cardboard carton with an inner liner like a mini bag-in-box, though it did include some plastic which can't be recycled.

Bob Lindo of Camel Valley offers something of an apocalyptic view: 'As time progresses and driven by climate change mitigation, glass bottles will become a curiosity and wine may become a thing of the past, like smoking.' Way to lighten the mood,

Bob. But here's someone who is hopeful about the future – Fergus Elias from Balfour. 'It certainly isn't all doom and gloom. Climate change and the ramifications for our industry are complex. But increased temperatures and growth of GDDs [growing degree days] will enable us to ripen our grapes to higher levels than even Essex can achieve, which will make for some fascinating wines.' Christian Seely of Coates & Seely is similarly optimistic. He doesn't think that in the foreseeable future England will be unable to make its signature sparkling wines. 'I am not too worried about the Douro Valley[87] either, and the Douro is much hotter and drier than England. We've done climate studies with some very serious consulting firms that showed ten, twenty, thirty, fifty years' prognoses and even then I am pretty confident that we are going to go on making great wines. In England, I am very far from being worried. I think the prognosis for the production of English Sparkling Wine and eventually still wine is extremely good for the next few decades.'

In the short term at least, it seems that English wines are only going to get better and better. But the big question – which I have delayed until the end of the book – is just how good are they now?

---

87 At Quinta do Noval, which is owned by AXA Millésimes, the French company where Seely is CEO

CHAPTER 19

Good for England

'I'll be honest, I expected to be more impressed'

Alder Yarrow

So are you going to tell the truth about English wine?' an Australian who works for a Champagne company asked me at the London Wine Fair when I told him about the book I was working on. 'What's that?' I asked. 'That it's no good,' he replied. He went on to explain how most of the sparkling wines were underripe, with off-the-scale acidity masked by bubbles and a load of sugar. 'It tastes alright cold, but you try it the next day when the fizz has died down and it just falls apart.' In 2022, American wine writer Alder Yarrow tried a line-up of English wines and was surprised by how low the quality was considering all the hype: 'I'll be honest, I expected to be more impressed. Maybe it was all my British friends and colleagues driving up my expectations, but I had the idea that some of these wines were going to blow me away. That definitely didn't happen,' he wrote.

I should make it clear that I don't agree with the Australian, and I disagreed with Yarrow on many of his judgements. But when researching this book, I did often find myself asking, 'Is this actually good or just good for England?' I even used the abbreviation GFE in my notes. My wife reminds me of it when I start enthusing about one wine or another over dinner. 'Is it good, or just GFE?'

In this final chapter, I'm going to look back at the wines I've tried in 2021 and 2022 and try to answer the thorny question: just how good is English wine? It's difficult to assess, since I tried most of these wines in anything but exam conditions: in a cellar with the winemaker, out in the vineyard with often charming viticulturalists, or over dinner. It's not easy to be objective under such circumstances.

The best place to start is with the sparkling wines. Overall quality is high, and on the whole the big names such as Nyetimber, Ridgeview, Balfour, Gusbourne and Wiston offer good quality throughout their ranges. But we're now beginning to see English producers take on top-end Champagne via £100–200 bottles. And though there's no doubt that these are exceptional wines, there is a feeling in the trade and the press that English producers are trying to run before they can walk. Tommy Grimshaw at Langham is scathing about the money some are charging. He tries to keep his non-vintage wines to around £30 a bottle with the more prestigious ones under £50, though he could easily charge more. I'm a big fan of Langham's wines and not just for the modest pricing. Other smaller producers doing great things are Westwell in Kent, Harrow & Hope in Buckinghamshire, Hundred Hills in Oxfordshire and

Everflyht in Sussex (I know, it sounds like a badminton racket). None of them is charging silly money.

It has to be said though, there are some boring bottles out there too; competently made wines that don't excite. And yes, Australian man, there are some green, underripe wines too, demonstrating that it is still not easy to get perfectly ripe grapes here. I'd like to see more producers making, like Nyetimber, non-vintage wines to smooth out the bad years. Some super-market own-labels frankly aren't very good and you'd be much better off going for a budget Champagne instead.

Overall though, if you avoid the very cheapest wines, I'd say you can buy English Sparkling Wine with confidence. The sweet spot for me is in the £30–50 bracket where, on the whole, I'd rather drink English over Champagne. So no, I think Australian man was largely wrong here. Regarding Charmat wines (those made using the Prosecco method), despite my scepticism about pricing, I've yet to drink one that I haven't enjoyed. I even liked the Harlot wines, especially the rosé. If they can come down in price a bit, then I think the style has a very bright future.

When it comes to still wines, things get a bit more com-plicated. I spoke to the head of buying at a major wine merchant who had tasted through the range of English still wines from Waitrose, a supermarket that specialises in English wine, and was surprised by the paucity of quality on show. There were only a couple of wines in the line-up that he would think about stocking, he said. I'll be honest – many if not most English still wines have that English pinch in the mouth: too much acidity, too little fruit and a touch of greenness. I can drink and even enjoy such wines, but they are very much GFE.

When I told the wine merchant about my shorthand, he thought I meant, 'Good for England!' as in, 'Well done, England'.

He thinks people buy English wine because they want to celebrate where they live. The feedback from his customers was that they liked the idea of drinking locally-made products and were prepared to pay for them. Yet he only stocks a very small range of English wines.

The most commonly seen English grape for still wine is bacchus. The English wine industry seems to have bet the farm on bacchus, which can in the right hands produce some wines of elegance and style but often just results in wines that taste grassy and one-dimensional. I will admit here that I may be in a minority as most of the general public seem to like it, judging by sales of Chapel Down Bacchus. You can see why growers like it, too: it yields well and produces an affordable, palatable wine every year.

But far more to my liking are blends of Germanic varieties like huxelrebe, ortega, bacchus and others, where the whole is greater than the sum of their parts. I appreciate that these blends may be a hard sell, as people seem to like buying wines based on grape varieties that they know, but I have particularly enjoyed such wines from Blackbook winery in London, Davenport in Sussex and New Hall in Essex, which makes a delicious Signature Blend of huxelrebe, reichensteiner and pinot blanc for £12.50 a bottle. Stunning value . . . for England.

For reasons that I hope are clear if you've come this far in the book, English still wines are always going to be expensive. It's all about yields: English producers will be lucky to get five tonnes of grapes per hectare; in contrast, some growers in Portugal get fifteen. So at the lower end of the market, a wine from a warmer climate is always going to be much cheaper than an English wine of similar quality.

Further up the scale, various English producers are getting serious about French varieties, in particular pinot gris, pinot blanc and chardonnay. Pinot gris especially seems to have a huge amount of potential, producing wines with the freshness of Northern Italy but the weight of Alsace. I've been particularly impressed with pinot gris from a new producer in Sussex called Artelium. Very much not just GFE.

As English wine pioneer Edward Hyams[88] wrote about English wine in 1965: 'Only by selling their wines at a price which has nothing to do with ordinary economic laws can they be made financially viable.' In other words, only by charging Burgundy prices can producers in England make money. Well, now it's happening. The success of Kit's Coty from Chapel Down was a game changer for English chardonnay. Not just the quality, but the price – around £30 a bottle. It's been joined by excellent wines from Oastbrook, Gutter & Stars, Danbury Ridge and others. These are exciting wines with price tags to match. Sadly, there are also some awful chardonnays out there. Warm vintages like 2018 and 2020 encourage producers of sparkling wines to bottle still chardonnays too, but they struggle in less good years like 2019 and 2021. Buy a cheap English chardonnay with caution.

I find English rosé immensely frustrating. There are too many Provence-coloured, Instagram-chasing examples which just taste grassy and insipid. With all the pinot noir around that's not quite ripe enough for a full-on red wine, you'd think that a cherry-red rosé could be England's signature style. Gusbourne even produced just such a wine in 2020 and it was gorgeous, one of my wines of the year. Sadly the style is not in

---

88 *Dionysus*

fashion, and the following year they went for a Provence-style version which wasn't a patch on it. Yet that happy medium between rosé and a red wine suits our climate. I tried quite a few examples, especially from the 2021 vintage, where the grapes weren't quite ripe enough to make a proper red, and I loved them. I'm hoping that the fashion for pale Provence-style wines fades quickly as I don't think it works in England. Rosés with colour please.

Which brings me on to red wines. Now this really is GFE territory. A good English pinot noir is always going to be expensive and rare. Some of the 2021 pinot noirs I tried were very green, though there were some ripe ones too. But if you've got pinot noir grapes, you're probably better off from a financial point of view making a sparkling wine out of them. The potential is definitely there though, as wines from Gusbourne, Balfour and, most of all, Danbury Ridge show. I haven't yet been convinced by the many pinot noir blends I have tried, which seek to bolster their naturally anaemic colour with a dose of rondo or regent. But I did really enjoy the Baron's Lane red blend from New Hall, which mixes pinot noir with all kinds of other things. It tastes like a good ordinary Beaujolais, and I mean that as a compliment. Drunk cold out of a tumbler makes a lot of sense – though my wife still thought it was just GFE.

So there we go. There's a lot of good English wine and much that you might find disappointing. Then again, there are indifferent wines everywhere. They've been making wine in Turkey for thousands of years and much of it still isn't very good. Furthermore, in other countries there are many wines that in strict quality terms are much too expensive, and yet still sell well to affluent local customers. I'm particularly thinking of California here, many of whose wines have English wine lovers scratching their heads like Alder Yarrow did about ours.

England might be a bit damp and chilly but it isn't even the most challenging place to make wines any more. There's now viticulture in Scandinavia with a strong local following. In Ningxia, the main winemaking region of China, it gets so cold that the vines have to be cut back for winter and buried in the ground to protect them – a very expensive operation given that many die anyway. Now that's winemaking in a cold climate. Makes a spot of mildew in Kent sound like a picnic.

England's wine, as befits an industry that is barely 50 years old, is very much a work in progress. Most New World countries which seemed to spring from nowhere in the eighties actually had very long-established unbroken traditions of making good quality wine. Penfolds in Australia was founded in 1844. Most English wineries are under 10 years old and there is still that undeniably challenging climate of ours to deal with. As Christian Seely of Hampshire's Coates & Seely says: 'We are making wines which can be compared with historic vineyards and that has been achieved in a remarkably short space of time. I am full of excitement and hope about what is going to happen in the next 15 years.'

Despite the rapid expansion of English winemaking, it's worth reminding ourselves just how small production in this country remains. England produces at most 15 million bottles a year in an unusually good vintage like 2018. Compare that with Champagne, which churns out around 300 million bottles. Or Prosecco, at 500 million bottles.

When I started writing this book, I had a particular image of the English wine industry: corporate, moneyed and, if I'm being honest, a bit boring. From the outside it all looks a bit 'branded gilet'. But as I discovered, it's not just about millionaires with a retirement plan. Some of the best wines are being made by young people with a garage, a few barrels and a lot of big ideas. Meanwhile, scratch the surface of the larger producers and there are plenty of forthright, opinionated people making fascinating wines that belie some of these companies' rather bland public images. When you consider how much this tiny industry has achieved in such a short space of time, with the sheer diversity and quality of wines we now produce, it's hard not to be as excited for the future as Seely. I hope I've managed to bring some of this to life and, most of all, to encourage you to try a few more of our wines from a cold climate.

GLOSSARY

ABV: alcohol by volume.

Biodynamics: A controversial system of agriculture devised by Rudolph Steiner. It's related to organics but with homoeopathic and spiritual elements. See Organic Growth, chapter 11, for more detail.

Blanc de Blancs: Sparkling wine terminology for a white wine made solely from white grapes, usually chardonnay.

Blanc de Noirs: A sparkling white wine made from black grapes, usually pinot noir and pinot meunier.

Botrytis: A fungus that shrivels and rots the grapes on the vine. Usually this is an undesirable thing but in certain circumstances when it affects ripe, undamaged grapes, the action of botrytis concentrates the sugars without ruining the taste of the grapes. The great sweet wines of Sauternes in France and Tokaji in Hungary are made with botrytis-affected grapes.

Charmat: cheaper method for making sparkling wines such as Prosecco. See Fizz Wars, chapter 8, for more detail.

Champagne method: See traditional method.

Chapitalisation: Adding sugar to a fermenting wine to boost alcohol levels.

Classic method: See traditional method.

Clone: A strain of a particular grape variety, ie. chardonnay.

Cuvée: The first pressing of juice, the best portion, is known as the cuvée. Also used as a term for different wines from the same producer, with the prestige cuvée being the top of the tree.

Disgorging: The process of removing the dead yeast cells from a bottle of sparkling wine so that the resulting wine is clear and bright.

Dosage: Sweetness added to a sparkling wine post-disgorgement. This could be in the form of grape juice, sweetened wine or even fortified wine.

Downy mildew: A fungal disease originally from North America that attacks the green parts of a vine and leaves cotton-like growth on the underside of the leaves. Severe attacks can stop the plant photosynthesising. The traditional treatment is Bordeaux mixture – copper sulphate and sulphur.

English Sparkling Wine: A legal term for a sparkling wine made using the traditional method in England. See Fizz Wars, chapter 8.

Geisenheim: The German wine school which has been hugely influential on English viticulture.

Glyphosate: A herbicide, the best-known brand of which is Round-Up, produced by Monsanto.

Growing degree days (GDD): A measure of how much warmth a plant will get for ripening in a certain climate. The number is calculated by taking the mean temperature in a day and taking away the base temperature (the temperature at which a plant will stop developing). All the days when the plant is developing are added together to give you the number of growing degree days. The mean growing degree days in southern England has increased from 850–900 during the period 1981–2000 to 1000–1050 from 1999–2018.

Hybrid: A vine made from a crossing of different species of vines, usually vitis vinifera with something else.

Master of Wine: A qualification issued by the UK-based Institute of Masters of Wine in oenology, viticulture and general knowledge of wine. It's extremely difficult to obtain and there are only around 400 MWs globally, working as consultants, merchants, winemakers and journalists.

Natural wine: An unregulated term, unlike organics, that signifies a wine made with minimum addition ie. cultured yeast strains, added enzymes etc. There's usually an implication that the grapes will be grown without synthetic chemicals but that is not always the case.

Non-vintage: A blend of different years, usually in sparkling wine, as opposed to vintage, which means the wine is all from one year.

Orange wine: A wine made from white grapes but using skin contact, as with a red wine, to impart flavour, colour and tannin from the skins.

Organics: Farming without synthetic chemicals. There are various bodies like the Soil Association in Britain which regulate this.

Pet nat: Short for petillant naturel, and a simple method for making fizz in a bottle. See Fizz Wars, chapter 8.

PDO: Protected Designation of Origin. An EU and now British legal designation for a foodstuff from a particular location that has to fit certain parameters in order to quality for the term.

Phylloxera: An aphid from America that eats the roots of vines and nearly wiped out the global wine industry in the late 19th and early 20th century. Vineyards were saved by grafting vitis vinifera vines onto resistant American rootstocks.

PiWi: A type of hybrid grape which has fungal resistance from the German *Pilzwiderstandsfähig*.

Powdery mildew (aka oidium): A fungal infection originally from North America which leaves grey powdery spores on the green parts of the vine. It stops the grapes growing properly and infected grapes will spoil the wine, making it taste mouldy. The traditional treatment is sulphur.

Residual sugar: The sweetness of a finished wine measured in grams per litre.

Remuage: The process used in sparkling wine to move the dead yeast cells into the neck of the bottle where they can be removed by disgorgement.

Rootstock: A usually non-vitis vinifera root onto which vitis vinifera vines are grafted to provide resistance against phylloxera.

Saignée: Method for making rosé, usually sparkling, wine. The word comes from the French for 'bled' and involves 'bleeding' a proportion of juice from a red wine that is still fermenting so it has some colour from the skins.

Traditional method: The method perfected in Champagne for secondary fermentation in bottles to produce bubbles. AKA Classic Method and Champagne Method. For a detailed look at how this works, see Fizz Wars, chapter 8.

Yield: The amount a vineyard produces, usually expressed in tonnes of grapes per hectare or hectolitres of wine per hectare.

Vitis vinifera: The Eurasian species of vines that probably originated in the Caucasus and now produces most of the world's wine.

BIBLIOGRAPHY

Barty-King, Hugh – *A Tradition of English Wine* (Oxford Illustrated, 1977)

Clarke, Oz – *English Wine* (Pavilion, 2022)

Chancellor, Edward – *The Price of Time* (Allen Lane, 2022)

Dallimore, Ed – *The Vineyards of Britain* (Fairlight, 2022)

Goode, Jamie – *Regenerative Viticulture* (Independently published, 2022)

Hyams, Edmund – *Dionysus: A Social History of Wine* (Sidgwick & Jackson, 1987)

Jeffreys, Henry – *Empire of Booze* (Unbound, 2016)

Jefford, Andrew – *Drinking with the Valkyries* (Academie du Vin, 2022)

Laithwaite, Tony – *Direct: The Story of Laithwaites* (Profile, 2019)

Ordish, George – *Vineyards in England and Wales* (Faber, 1977)

Pearkes, Gillian – *Vinegrowing in Britain* (The Blackwell Press, 1982)

Rose, Anthony – *Fizz* (Infinite Ideas, 2021)

Sagres, Liz – *A Celebration of English Wine* (The Crowood Press, 2018)

Skelton, Stephen – *Wine of Great Britain* (Infinite Ideas, 2019)

Stevenson, Tom & Avellan, Essi – *Christie's World Encyclopedia of Champagne and Sparkling Wine* (Bloomsbury, 2019)

ACKNOWLEDGEMENTS

The vast majority of the research for this trip was undertaken at my own expense, but I did go on three trips which were paid for by third parties: to Roebuck Estate in Sussex, Ridgeview, also in Sussex and to multiple wineries as part of a tour organised by Great Sussex Way tourist body.

I'd like to thank everyone who took part in this book. Everyone, or almost everyone, has been incredibly welcoming, putting up with my silly questions and letting me try lots of wine. Certain people, however, were particularly helpful and I'd like to thank especially Charlie Holland at Gusbourne, Simon and Mardi Roberts at Ridgeview, Fergus Elias at Balfour, John Atkinson at Danbury, Adrian Pike at Westwell, Chris Wilson from Gutters & Stars, Father Christopher Lindlar, Jerome Moisan, and Sandy and the late Stuart Moss. I'd also like to thank the various people who read early drafts of the book and gave me their feedback both positive and negative. Also thank you to my employer Master of Malt for letting me work three days a week; without their flexibility I don't see how this book could have been completed.

At Atlantic Books thank you to Kate Ballard for being the model of calm as I completely rethought the book, twice, and handed it in a month late, plus Will Atkinson, Karen Duffy and Derek Wyatt, whose idea it was in the first place. Thank you also to Guy Woodward for his thoughtful and sympathetic edit and to my agent Jo Cantello.

Most of all, thank you to my wife Misti and daughters Edith and Helena for putting up with me disappearing for long periods of time at unhelpful moments. Writing a book is hard for the author but doubly hard for the family, so thank you and sorry. I promise I won't write another book that requires so much research.

INDEX

Henry Jeffreys studied English and Classical Literature at Leeds University. He worked in the wine trade and publishing before becoming a freelance writer and broadcaster. He was wine critic for *The Lady*, and his work has appeared in *Spectator* magazine, the *Guardian*, the *Oldie* and *BBC Good Food* magazine. He has been on BBC Radio 4, Radio 5 and Monocle Radio, and featured on BBC 2's *Inside the Factory* (2020). He is the author of the award-winning *Empire of Booze: British History Through the Bottom of a Glass* (2017), *The Home Bar* (2018) and *The Cocktail Dictionary* (2020), and in 2022 was awarded Fortnum & Mason Drink Writer of the Year. He is currently features editor for the Master of Malt drinks blog and drinks writer for *The Critic* magazine. He lives in Faversham, Kent with his wife and two children.